P. E. REVIEW

Coordinator: C. V. Chelapati

Urban Storm
Drainage Management

CIVIL ENGINEERING

A Series of Textbooks and Reference Books

Editors

ALFRED C. INGERSOLL
Associate Dean, Continuing Education
University of California, Los Angeles
Los Angeles, California

MILO S. KETCHUM
KKBNA Engineers Inc.
Denver, Colorado

KENNETH N. DERUCHER
Department of Civil Engineering
Stevens Institute of Technology
Castle Point Station
Hoboken, New Jersey

In Preparation

Urban Storm Drainage Management

JOHN R. SHEAFFER

Sheaffer & Roland, Inc.
Chicago, Illinois

KENNETH R. WRIGHT

Wright-McLaughlin Engineers
Denver, Colorado

with

WILLIAM C. TAGGART

Wright-McLaughlin Engineers
Denver, Colorado

RUTH M. WRIGHT

Colorado Legislature
Denver, Colorado

MARCEL DEKKER, INC.   New York and Basel

Library of Congress Cataloging in Publication Data
Main entry under title:

Urban storm drainage management.

   (Civil engineering ; 5)
   Includes bibliographical references and index.
   1. Flood damage prevention. 2. Urban runoff.
I. Sheaffer, John Richard. II. Series.
TA1.C4523 vol. 5 [TC530] 624s [363.3'49372] 82-2500
IBSN 0-8247-1351-6   AACR2

MARCEL DEKKER, INC.
270 Madison Avenue, New York, New York 10016

Current printing (last digit):
10 9 8 7 6 5 4 3 2 1
PRINTED IN THE UNITED STATES OF AMERICA

Foreword

For a long time, the disposal of storm water in modernized urban areas has tended to give primary attention to the costs and benefits of different sizes and layouts of drains in relation to postulated precipitation. The effort has been to transport the flows from intense or protracted rainfall by means that seemed economical in the light of expected losses when flows exceeded system capacity. Much analysis was directed at estimating the damages averted or suffered as a result of various return intervals and volumes of flow considered in the design.

A sometimes secondary set of problems arose over the issue of the extent to which storm drainage and sanitary sewer drainage should be separated. Both of these issues continue to claim central attention, but a number of other considerations are entering into the management process. This book serves to outline practices in moving from rainfall analysis to system design, but it incorporates study of the ways in which a truly integrated program of storm drainage management might be achieved. The broadened effort requires an examination of unified floodplain management, of the effects of urbanization on flood flows, of alternative methods of water storage, and of the links between storm-flow disposal and water-quality planning.

Evolving views of floodplain management now encompass storm drainage, but in a framework which includes land-use planning and regulation, land acquisition, flood-warning systems, flood proofing of buildings, and insurance. To make a comparative analysis of these numerous alternatives in a variety of combinations requires knowledge going beyond the more conventional engineering. The emphasis placed on the several alternatives is bound to change as national policies develop, for example, with respect to national support of land acquisition or of flood loss insurance. Nevertheless, recognition of the interrelationship of these several types of measures in programs to cope with storm flows is strong and is unlikely to decrease.

Mounting evidence as to effects of upstream urbanization on the timing and magnitude of high-frequency flows makes it essential to plan any flood-loss reduction program with an eye to those alterations in the flood hydrograph. All too often, downstream floodplain measures are designed without anticipating the consequences of storm drainage elsewhere in the drainage area.

A relatively new and significant aspect of drainage design is the conscious provision of detention storage in streets, parking lots, and behind road fills. This method of supplementing natural storage in some reaches may be an integral element in a system that accelerates flows in other reaches.

iii

The consideration of water quality in the storm runoff is encouraged in part by public policies to reduce or prevent water pollution, but it also is stimulated by the broader view of drainage as a component in watershed and floodplain management. Flows can no longer be regarded solely as volumetric qualities. They need to be seen as differing greatly in the materials—natural and man-made—they carry in suspension or solution.

In all four of these ways this volume strives for a genuinely comprehensive approach. It goes beyond the more conventional engineering approaches and draws upon insights and methods from other disciplines as well. Significantly, its authors are engineers, a geographer, and a lawyer serving as legislator. They draw from experience inside and outside the United States. In one sense, they make the design process more complex by examining a large number of complementary techniques. In another sense, they make storm drainage design more coherent by showing its place in the realities of land and water use in the affected watersheds.

Gilbert F. White
Gustavson Professor Emeritus of Geography
University of Colorado at Boulder

Preface

Every urban area has one or more public agencies designated to serve as the major drainage and floodplain management authority. A number of other public and private interests, including various municipalities and land developers, also share in this responsibility. Accordingly, this book has been prepared to help the various interests achieve a unified conceptual approach. Hopefully, use of the book will lead to improved drainage and floodplain management in urban areas.

There are two basic elements in a comprehensive floodplain management program. First, there are the *preventive* measures which focus on floodplain regulations in their broadest context. The preventive elements will not allow the existing flood loss potential to increase. The second element is *corrective* in nature. The corrective approach seeks to mitigate the existing flood losses which result from unwise development of floodplains or underdrained areas. This unwise development may have taken place before there was an awareness of the flood hazard or before preventive elements were enacted to control such development.

The preventive and corrective approaches presented in this book are multipurpose and ideally should be implemented by multiple means. When properly planned, floodplain areas provide opportunities to help improve the quality of life. In many locations watercourses radiate like spokes through urban areas into the countryside, linking rural and urban places. Urban networks and recreational uses can be integrated into the floodplain areas. Such networks may include utility lines and transport facilities. Where appropriate, floodplains in public ownership could be developed as linear parks that provide recreational facilities including trails for hiking, cycling, and horseback riding. Any such development must, however, have regard to overall metropolitan development policies.

This book contains four primary areas of guidance and information. Each of these can be applied to both preventive and corrective main drainage or flood control programs:

Goals: Here desirable goals, objectives, principles, policies, and criteria of drainage and flood control are presented.

Legal aspects: This overview summarizes some of the legal opportunities and constraints involved in implementing the approaches and techniques of drainage and flood control management under present legislation.

Strategic drainage planning process: Planning for drainage or flood control, whether preventive or corrective, should be an orderly and consistent process. When all interests employ the same planning process, a unified conceptual approach is achieved.

Basic technical criteria and information: Basic information on rainfall and runoff is presented which serves as the basis of design for both preventive and corrective programs. The section on floodplain delineation allows spatial definition of the problem. The floodplain regulation chapter outlines needs and criteria relating to the legal control of floodplain development. The chapters on non-structural planning and man-made storage illustrate both preventive and corrective tools. The last chapter, on water quality, presents background information regarding the developing science of runoff water quality.

The preparation of this book was aided by inputs from the staffs of Sheaffer & Roland, Inc., and Wright-McLaughlin Engineers, Inc. Their experience in drainage evaluations and solutions and floodplain management program formulation and implementation provided the background and experience upon which the book is based. Special recognition is due to Dr. A. C. Ingersoll, Associate Dean, University of California Los Angeles, for his critical review and evaluation of the manuscript.

John R. Sheaffer
Kenneth R. Wright
William C. Taggart
Ruth M. Wright

Contents

Contents

Urban Storm
Drainage Management

1

Statement of Goals, Objectives, Principles, Policies, and Criteria

1.1 INTRODUCTION

Planning Strategy

There are many demands on the land and water resources of an urban region. The demands are associated with efforts to achieve a variety of objectives such as economic development, regional development, transport, social well-being, and environmental quality. Because these resources are limited and the demands are not, these objectives compete with one another. In terms of drainage, competing objectives must be considered and reconciled through a formal planning strategy. A planning strategy contains:

> *Goals.* Broad general purposes and programs
> *Objectives.* Specific end results desired as program results
> *Principles.* Conceptual framework which governs overall effort
> *Criteria.* Specific operational requirements

Within this context, demands for the land and water resources can be ordered in terms of their ability to achieve desired goals and objectives. This is a powerful device that can be used in the decision-making process. Drainage must be viewed as one of many issues affecting the use of land in the metropolitan area. While floodplain areas can provide recreational opportunities, not all recreation areas should be dependent on drainage considerations.

Drainage basins are convenient units for water resources management purposes. Within the boundary of each drainage basin, a system of watercourses has evolved which is specifically related to the physical and hydrologic conditions. The watercourses, and the floodplains developed through periodic inundations, are the primary areas of consideration in drainage basin management. However, to mitigate flood losses, control erosion, manage sedimentation, and abate water pollution, it is necessary to formulate management policies not only for the watercourses and floodplains, but also for all parts of the drainage basin.

Floodplain Description

A floodplain is generally a smooth or relatively flat area bordering a stream which is built of sediments carried by the stream, but may also include part of the steeper valley slopes. It is periodically inundated in part or totally, depending on the magnitude of the floods. The probability of inundation is an important concept in drainage basin management. The floodplain can be defined as the area subject to flooding by the flood of record or the maximum probable flood. For this book, the usual reference is to the floodplain resulting from the flood which is estimated to have a 1 percent chance of occurring in any one year (the 100-year flood).

Drainage Agency Responsibilities

A drainage agency is in a unique position to assist in the formulation of goals and objectives which, if achieved, will result in the community being able to obtain the maximum possible benefits from its watercourses and their environs. In formulating these goals and objectives, it is necessary that there be a sensitivity to the competing demands on resources within the drainage basin. Drainage or flood control should be developed within the context of overall strategic objectives for planning the development of the region. By clearly articulating goals and objectives, principles, policies, and criteria can be formulated to assure that drainage planning is characterized by goals achievement.

1.2 GOALS

Drainage or flood control should be viewed as an integral part of the comprehensive planning process. It is a subsystem of a larger and more comprehensive urban system.

A primary goal is to have a unified conceptual program for drainage and flood control because such a program can mitigate future flood damages while systematically reducing existing flood damage through comprehensive floodplain management. Where undeveloped floodplains exist, land uses should be controlled to prevent development that would result in increased flood losses. Floodplains left open are community assets. They can provide the most economical and socially advantageous flood loss mitigation measure. Existing flood problems can be mitigated by applying the proper combination of nonstructural and structural measures.

A regional drainage agency can play a leadership role by providing technical information and by encouraging municipalities to make land use adjustments to achieve a broad range of benefits while preventing the unwise use of floodplains. The drainage and flood control program should be coordinated with the water supply, sewerage, land use planning, transportation, and open space programs.

1.3 OBJECTIVES

Within the context of overall development of a region, drainage programs should generally be governed by the following objectives:

> To retain nonurbanized floodplains in a condition that minimizes interference with floodwater conveyance, floodwater storage, special important aquatic and terrestrial ecosystems, and ground/surface water interface
>
> To reduce the exposure of people and property to flood hazard
>
> To systematically reduce the existing level of flood damages
>
> To ensure that corrective works are consistent with the overall goals and objectives of the region
>
> To minimize erosion and sedimentation problems

To protect environmental quality and social well-being
To improve the usefulness of floodplains which are developed as active
and passive recreational areas

1.4 PRINCIPLES

The basic principles underlying the goals and objectives for drainage and
flood control include the following:

1. The drainage system is part of a larger environmental system. The
drainage system is a part of a larger interrelated comprehensive urban system.
The drainage system can be managed as simply a support system for an urban
area or it can be managed in a way that will assist efforts to achieve a broad
range of goals and objectives. In the latter sense it is a means to an end not
an end in itself.

Urbanization has the potential to increase both the volume and rate of
stormwater runoff. The influence of planned new development within a drainage basin must be analyzed and adjustments made to minimize the creation of
flood problems. Local and regional goals help define the drainage works prescribed for a watercourse.

2. Floodplains are natural storage areas. The floodplain is nature's
prescribed and natural easement along a watercourse. The primary natural
functions of each watercourse and its associated floodplain are the collection,
storage, and transmission of stormwater runoff. These functions cannot be subordinated to any other use of the floodplain without costly compensatory control measures. Within those constraints, the floodplains have the potential
to help improve water quality and air quality, provide open space, preserve
important ecosystems, and accommodate properly planned urban network
systems.

3. Stormwaters require space. Stormwater management is a time-related
space allocation problem. Water cannot be compressed, and if natural storage
is reduced by urban or other land use practices, without appropriate compensatory measures, then additional space will be claimed by the floodwaters at
some other location(s).

4. Stormwaters have potential uses. Stormwater can be viewed as a
resource out of place. In such cases storage of stormwater is the first step in
a program to make use of the resource. These storage areas can be designed
and operated to provide aesthetic amenities and recreational space. The stored
water may have the potential to be used for irrigation, groundwater recharge,
low-flow augmentation, and industrial water supplies.

5. Water pollution control measures are an essential feature. Water pollution control is essential to a realization of the potential benefits to be derived
from watercourses and floodplains. Pollution control measures, which deal with
both point and nonpoint discharges, are an integral part of a drainage and
flood control program. The 1977 Clean Water Act mandates the identification of
open space and recreation opportunities that can be expected to result from
improved water quality.

1.5 POLICIES

To achieve the stated goals and objectives of drainage basin management, the
following policies should be implemented:

1. Establish criteria. A drainage agency should establish and publish criteria for drainage and flood control planning and design. Guidance relative to construction, operation, and maintenance of urban drainage systems should be provided also. The agency should encourage the adoption of the criteria by all relevant public and private drainage interests. Such criteria should be periodically reviewed and revised in the light of new knowledge, changing circumstances, and adjustments in overall comprehensive goals and objectives.

2. Establish drainage plans. The drainage agency, in conjunction with affected local governments, should prepare plans for each watercourse. These plans should identify associated land uses. Inputs should be elicited from appropriate interest groups during the planning process. The plans should be placed on exhibition for public comment prior to their adoption.

3. Plan revisions. Plans for drainage basins should be periodically reviewed and revised in the light of new knowledge, changing circumstances, and adjustments in overall comprehensive goals and objectives. Unless otherwise determined, such reviews should be at intervals of approximately 10 years.

4. Plan compatibility. The cooperation of local governments and other relevant drainage interests, including the land development industry, should be sought to coordinate individual development and drainage schemes with the plans.

5. Possible strategies. Consideration should be given to a full range of preventive and corrective approaches, including the following:

Delineation of floodplains
Control of floodplain land uses
Acquisition of selected floodplains
Floodplain information and education
Flood forecasts and emergency measures
Flood proofing
Flood insurance
Restriction of the extension of water and sewer facilities in floodplains
Detention (retardation) and retention of urban stormwater runoff
Construction of flood control and drainage works

The combination of strategies should be tailored to a specific site and will balance engineering, economic, environmental, water quality, and social factors in relationship to stated comprehensive goals and objectives.

Multiple-objective floodplain management requires multipurpose planning. Where multipurpose benefits will result from the implementation of the drainage policy, funds from other appropriate sources should be sought to supplement the drainage agency's funds.

6. One percent design flood. In general, proposals for both structural and nonstructural projects will be designed to accommodate likely discharge arising from the appropriate critical-duration rainfalls of 1 percent probability. (The 1 percent probability rainfall—or 1 in 100-year event—is that having a 1 percent probability of being exceeded in any year.)

7. Flood proofing. The National Flood Insurance Program has the responsibility of delineating the special flood hazard areas. The floodplains are delineated on the basis of the 1 percent flood (100-year flood). The 1 percent flood is computed by using synthetic hydrology based on rainfall-runoff relationships or by statistical analysis of flood records where these are reliable and long-term; significant alteration of the watershed is not expected. The probability of experiencing a 100-year flood during a 100-year interval is 63.4 percent.

Flood insurance rate maps also delineate the 500-year floodplain. The probability of experiencing this higher flood within a 100-year interval is 18.1 percent.

8. *Preventive actions.* Proposed development within the special flood hazard area should be adjusted to be compatible with the flood hazard. The use of the floodplain will be viewed within the context of a regional land use plan. Also to be considered are the range of natural benefits provided by floodplains.

9. *Corrective actions.* It is necessary to develop and implement drainage and flood control plans that should mitigate existing drainage problems. Such plans should be coordinated with comprehensive goals and objectives and will consider a combination of structural and nonstructural measures.

10. *Pollution control.* Pollution control programs should be integrated into the drainage and flood control program.

2

Legal Aspects of Drainage

2.1 INTRODUCTION

A storm moves in over a basin. The rain hits the earth—some of the water percolates into the ground, some of it flows overland in a diffused manner, collecting in depressions and swales, gathering in gullies, eventually to flow into and become creeks and streams. If the storm is of a great magnitude, the water cannot be contained within the banks of the creeks and streams, so the water spreads out over the floodplain—the natural path it has created for itself over geologic time.

With the advent of man, a storm still moves in over the basin and the waters move downhill. But now there are changes in the natural topography. Depressions are filled in. The land is made impermeable by streets, parking lots, and rooftops, resulting in less water percolating into the ground. Streets and storm sewers collect the water so that more water with greater velocity is discharged onto lower lands, or discharged in a different location. Embankments and dikes are built which divert the course of floodwaters and reduce natural detention. Roads and bridges constrict the flows, causing waters to back up and flood lands which would not have been flooded, or would have been flooded to a lesser extent. Rivers are straightened and channelized, which speed up the flow and cause greater impact downstream.

These changed conditions can cause injury greater than formerly and spawn lawsuits requesting damages for the injury and/or injunctions to prevent injury. In addition, as government steps in to attempt to manage surface waters, watercourses, and floodplains by constructing facilities or by the use of police power (drainage ordinances, subdivision regulations, floodplain zoning, and other techniques), a host of other legal confrontations arise.

It is essential that governmental entities, planners, and engineers work within a legal framework for a number of reasons: to avoid legal obstacles to implementation of plans and designs; to minimize potential liability; and to be alerted to opportunities that may be available. Legal principles may sometimes even become the determining factors in deciding among viable engineering alternatives. While it is always essential to ascertain the specific state and local laws operative in a particular jurisdiction where the work is being done, various aspects of stormwater law can be presented here in a general fashion. These include the court-made law of watercourses and surface water, potential liability, floodplain management, water pollution control, and a drainage fee concept.

7

2.2 LAW OF WATERCOURSES AND SURFACE WATERS

Distinction between Watercourses and Surface Waters

The law of watercourses and surface waters is essentially court-made, that is,
it developed when lawsuits were brought before the courts to decide the
rights, duties, and liabilities between private landowners in drainage and
flooding cases. An early distinction was made, and is still being made, between water flowing in a watercourse and surface waters. Surface waters
are waters which run in a diffused manner overland, or in depressions or
swales, while a watercourse has definite banks and bed. Floodwaters which
overflow the banks of a watercourse and follow the course of a stream to its
natural outlet, or which upon subsidence return to the stream, are also held
to be governed by the law of watercourses. Floodwaters which entirely lose
their connection with a lake or stream, however, and spread out over the
adjoining countryside and settle in low places are probably governed by the
law of surface waters.

   Where state courts have adopted surface water rules that are incompatible with their watercourse rules, a decision regarding liability may hinge
totally on the category (watercourse waters or surface waters) into which the
errant waters are placed [*MacManus v. Otis*, 61 Calif.App.2d 432, 143 P.2d
380 (1944)].* The problem, of course, is that the court must decide at what
geographical point a swale or gully suddenly becomes a watercourse, at which
point there is a switch in the rules resulting in an opposite decision. Not
only is this point sometimes difficult to ascertain, but it is also inequitable
since stormwaters are still part of the same hydrologic system. Liability for
actions should not hinge on so tenuous a basis.

   On the other hand, where the court has adopted compatible surface
water and watercourse rules, the dilemma and inequities are avoided [*Davis v.
Frey*, 14 Okla. 340, 78 P. 180 (1904); *Chicago, R.I. & P. Ry. v. Groves*, 20
Okla. 101, 93 P. 755 (1908)].

Law of Watercourses

Watercourse law is based on the rights and duties established between riparian
property owners. The term *riparian* means pertaining to or situated on the
bank of a river or lake, and includes land along a seasonal wash or arroyo.
The fundamental principle of the riparian system is that each riparian owner
has an equal right to make a reasonable use of the water of a stream, subject
to the equal right of the other riparian owners to do likewise. Therefore, a
riparian owner must exercise his (or her) rights in a reasonable manner so as
not to interfere unnecessarily with the corresponding rights of others.

   Regarding flood flows, a riparian owner has a right to protect his land
from ordinary floodwaters, but only if he does not unnecessarily injure other
riparian owners thereby. A rural landowner who built an embankment along a
stream which in flood times protected his land but cast more water onto
another's land, causing damage, was liable for such damage [*Beetison v.
Ballou*, 153 Nebr. 360, 44 N.W.2d 721 (1950)]. Where a town built a dike to
protect itself from recurring floodwaters, which in turn were diverted onto a
neighboring farm, the court required removal of the dike [*Town of Jefferson
v. Hicks*, 23 Okla. 684, 102 P. 79 (1909)]. The principle, then, is that

---

*All cases cited in this chapter are listed under "Cases" at the end of the
chapter. Cases are cited in full in the text only at first mention.

owners of lands bordering a natural watercourse are entitled to have its water, whether within its banks or in its flood channel, to flow in its accustomed path, and no one has the lawful right by diversions or obstructions to interfere with its accustomed flow to the damage of another. However, a distinction has been made between "ordinary" and "extraordinary" floods. The general rule has been that riparian owners may protect themselves from extraordinary, unprecedented floods without liability, unless negligence is involved. This issue then becomes: Was the flood ordinary or extraordinary? This issue is discussed under "Ordinary and Extraordinary Floods."

Law of Surface Waters

There are three basic doctrines which courts have adopted regarding surface waters. The two traditional ones based on property law are the *common enemy rule* and the *civil law rule*. A third, based on tort law, has evolved in recent years and is called the *reasonable use rule*.

As originally conceived, under the common enemy rule a landowner could do anything he pleased with surface waters regardless of the harm it might do to others. The upper landowner could divert or drain surface waters onto the lower land, and the lower landowner could put up a barrier even if it flooded the upper property. Since the water must go somewhere, this would result inevitably in contests in engineering where might makes right. Therefore, some courts have modified the strict rule, resulting in a "modern common enemy rule" which gives the landowners the right to obstruct or divert surface waters, but only when such obstruction or diversion is incidental to the ordinary use, improvements, or protection of their land and is done without malice or negligence.

Where one finds that a state court has adopted the common enemy rule, one must be cautious about jumping to conclusions. One must analyze how the courts actually apply the rule. In Montana, for example, the courts have so modified the rule that the decisions now rest on the overriding principle that one must so use one's property as not to infringe on the rights of others [*State Highway Commission v. Biastoch Meats, Inc.*, 145 Mont. 261, 400 P.2d 274 (1965)]. Oklahoma courts have modified their common enemy rule by the *rule of reason*. This rule results in liability for landowners who alter natural runoff if such alterations cause injury to others [*Chicago, R. I. & P. Ry. Co. v. Taylor*, 173 Okla. 454, 49 P.2d 721 (1935)]. An Iowa court has put it this way:

> We recognize the fact ... that surface water ... is a common enemy, which each landowner may reasonably get rid of in the best manner possible, but in relieving himself he must respect the rights of his neighbors, and cannot be justified by an act having the direct tendency and effect to make that enemy less dangerous to himself and more dangerous to his neighbor. [*Livingston v. McDonald*, 21 Iowa at 174 (1866)]

Under the civil law rule, the upper landowner has an easement for the natural drainage from his property over the lower property, and the lower landowner must take such water. The upper land is the dominant estate; and lower land is the servient estate. The key word is *natural*, meaning those waters which flowed from the land before alteration or development. It was felt that in its pure form, this rule would prohibit development. Many jurisdictions therefore interpret the rule to allow the upper owner to increase the natural flow to some degree, especially in an agricultural setting. The degree has been limited, for example, by balancing benefits and harms, or requiring that the landowner not act unreasonably or negligently. Colorado's version of the civil law rule is that natural drainage conditions may be altered by an upper proprietor provided the water is not sent down in a manner or quantity to do more harm than formerly [*Hankins v. Borland*, 163 Colo. 575, 431 P.2d 1007 (1967)].

The reasonable use rule is based on tort rather than on property law principles. A general tort principle is that you must use your property in a manner that does not injure the property of another. This doctrine permits the courts simply to recognize the right of each owner to deal with his property as he wishes so long as he acts reasonably under all circumstances. A disadvantage is the uncertainty of knowing in advance what a court might find reasonable or unreasonable under a particular set of facts.

Another area of uncertainty is that while legislation affects future actions (prospective), court decisions are retroactive. The court may find a defendant liable for breach of a duty which had not been articulated or established prior to the lawsuit. In addition, a court may decide that the old doctrine no longer suits modern-day problems and adopt a new doctrine, applying it to the case at hand. An example of this is the Wisconsin Supreme Court decision to adopt the reasonable use doctrine to supersede the common enemy rule in *State v. Deetz*, 66 Wisc.2d 1, 224 N.W.2d 407 (1974).

Ordinary and Extraordinary Floods

Whether or not a defendant was liable for his or her actions has often depended upon whether the flood was found to be an extraordinary rather than an ordinary flood. The extraordinary flood was held to be an act of God and a defense against liability. Even early on, however, some courts refused to absolve the defendant when damages occurred due to the act of God *plus* the defendant's negligence, where such negligence was the proximate cause of the injury [*Frederick v. Hale*, 42 Mont. 153, 112 P. 70 (1910)]. With the advances in hydrology and meteorology over the past decades, courts are now able to base their decisions on scientific data presented to them by expert witnesses, resulting in a rejection of the act-of-God defense. An example is *Barr v. Game, Fish and Parks Commission*, 30 Colo.App. 482, 497 P.2d 340 (1972). A state agency had built a dam and spillway. When a large storm occurred, the inadequate spillway caused water to flow onto the plaintiff's land. Evidence was presented that a competent meteorologist using modern techniques could have determined that the maximum probable flood for that basin would have a magnitude of 200,000 ft^3/sec. The spillway could handle only 4500 ft^3/sec. The actual flood flows had been 158,000 ft^3/sec. Had the scientific facts been ascertained and the spillway designed accordingly, it could have passed the flood. Under these circumstances the state agency could not avoid liability under the act-of-God defense. This case, then, also established the maximum probable flood as the criterion for spillways built by the state, and is an example of retroactive application of the court-established criterion.

Courts also use adjectives such as *unprecedented* and *unanticipated*. With scientists now regularly assigning a statistical probability to a given magnitude of storm/flood, with the 1 percent flood used by the National Flood Insurance Program, and with the Army Corps of Engineers designing for the Standard Project Flood, the question arises whether there is still a flood which can be termed "unanticipated." In cases where the court may feel that the defendant's actions were reasonable and nonnegligent in the particular circumstances, however, it still uses the unprecedented flood defense. In *City of Pascagoula v. Rayburn*, 320 So.2d 378 (Miss. 1975), a city maintained a natural waterway as a drainage ditch. A storm "in excess of 100 years" caused water to back up and damage plaintiff's property. The court found that the city had properly maintained the drainageway and held the city not liable because the storm was unprecedented.

See also *Oklahoma City v. Rose*, 176 Okla. 607, 56 P.2d 775 (1936). Here the city had built its facilities to handle more than double the flood of record, after consulting with and following the advice of nationally known authorities. The facilities were unable to pass a flood of even greater magnitude, causing damage to upstream properties. The court held that the city was not negligent and that the damages were due solely to an act of God.

However, where the flow was increased by upstream urbanization, and the city could have avoided the damage by simply replacing an inadequate culvert in a highway, the court held that the flood was not extraordinary but could have been anticipated [*Cabral v. City and County of Honolulu*, 32 Hawaii 872 (1933)].

Some courts hold that the flood of record is an ordinary flood, and where a new flood of record occurs, there is a duty to respond in a timely fashion to the new flooding conditions [*Missouri, K & T. Ry. Co. v. Johnson*, 34 Okla. 582, 126 P. 567 (1912)]. In the cited case the railroad company had built a bridge and culvert across a narrow valley just below the plaintiff's property; they were adequate for conditions known at the time. In May, 5 years later, the facilities caused water from a large flood to back up onto plaintiff's property 8 ft deep. Then in October of that same year an even greater flood put 12 ft of water onto the property. The court held that a "new standard of obligation" had been created by the May flood and that the railroad company had a duty to meet the new conditions. Having failed to do so, it was liable.

Conclusions

In terms of an actual lawsuit, the legal distinctions between watercourses and surface waters, and among the three rules of surface waters, may decide the outcome of the case. The defendant's attorney will argue whatever defenses these legal principles may provide in that jurisdiction. In terms of drainage *planning*, however, the potential for lawsuits and liability should be minimized. A good rule to follow in any jurisdiction is to maintain the natural runoff conditions as much as possible so that the facilities or regulatory scheme do not increase the potential harm to property and lives. It would also be wise for the appropriate governmental entity to establish reasonable criteria as to which flood magnitudes are to be recognized, controlled, regulated, and designed for in the long-term public interest for a particular purpose.

The 100-year (1 percent) flood may be appropriate for some purposes. However, because of the disastrous consequences of failure, an embankment across a channel to impound a flood may have to be designed to prevent failure under *any* probability. On the other hand, drainage facilities which are built for convenience purposes rather than protection of life and property can be designed for less stringent criteria. However, reasonably anticipated upstream urbanization should be incorporated into the design. A basinwide plan which takes all of these factors into account and manages stormwater in a prudent way will be most successful from both an engineering and legal point of view.

2.3 POTENTIAL LIABILITY

Private Liability

In the preceding section the rights and duties between private parties in
the drainage and flooding context were presented. While having their origin
in a rural setting, these principles are applied to the urban setting as well.
The upper landowner may be a developer who is subdividing a farm, constructing homes, or building a shopping center. Such urbanization usually
results in the former agricultural land being covered by impermeable surfaces such as homes, carports, sidewalks, streets, and parking lots. In addition, drain tiles and storm sewers are installed. The result is a substantial
increase in runoff during storm events and may even cause constant low
flows of surface water which formerly would have percolated through the
ground. Unless drainage planning is incorporated at the outset, which
either maintains the predevelopment flows or facilities are designed to take the
additional flows to a suitable point of discharge, damages can occur to lower
properties and liability ensue. *Armstrong v. Francis Corporation*, 20 N.J.
320, 120 A.2d 4 (1956) was such a case. The developer was held liable for
adding more land to the basin (thereby causing additional runoff which had
historically flowed elsewhere), for increasing runoff, and for installing
underdrains to collect and discharge groundwater, all of which caused
damage to downstream owners.

In a similar situation the court in *Breiner v. C & P Home Builders, Inc.*,
398 F.Supp. 250 (E.D.Pa. 1975), affirmed in part and reversed in part by
536 F.2d 27 (3rd Cir. 1976), held that the developer had unreasonably
increased the surface flows. Damages were awarded on the basis of cost to
cure the situation. In *Godlin v. Hockett*, 272 P.2d 389 (Okla. 1954) a subdivider dredged and deepened a creek and built a dike to protect his subdivision from flooding. His actions diverted floodwaters onto other riparian
properties in increased volume and greater depth. He was liable. It should
be pointed out that liability in these cases was based on violations of the
common law of surface waters and watercourses, independent of any governmental subdivision regulations with which the developer may or may not have
complied.

Liability of Public Entities

No Duty to Provide Drainage Facilities
Under common law there is no duty to provide for drainage and flood control
facilities [*Oklahoma City v. Evans*, 173 Okla. 586, 50 P.2d 234 (1935)].
Having provided such facilities either voluntarily or by statutory mandate,
however, public entities are treated much like private parties when their
facilities cause damage.

Liability for Negligence
Public entities are generally held liable for their own negligence. For example, in *City of Vicksburg v. Porterfield*, 164 Miss. 581, 145 So. 355 (1933),
the city built an embankment across a swale with a drain under the street.
The drain became clogged and heavy rains flooded plaintiff's property. The
court held that the city had a duty to provide for such floods as may reasonably be expected and had failed to do so. Therefore, the city was liable.

Strict Liability
In *City of Houston v. Wall*, 207 S.W.2d 664 (Tex. Civ.App. 1947) the city
built drainage works which were intended to and did in fact accelerate the

flow of rainfall into a creek abutting plaintiff's property. This acceleration caused flooding in a heavy, but not unprecedented, storm. In such a situation the city was strictly liable; that is, no negligence had to be shown.

Liability for Violation of Watercourse Law
Town of Jefferson v. Hicks and *Oklahoma Ry. Co. v. Boyd*, 140 Okla. 45, 282 P. 157 (1929) are examples of cases where public entities were found liable for violations of watercourse law.

Liability for Violation of Surface Water Rules
The civil law rule was invoked in *Dayley v. City of Burley*, 524 P.2d 1073, 96 Idaho 101 (1974). When the city expanded into a rural area, the former ability of the agricultural land to absorb water was lost. The city built curbs, gutters, and storm drains. The court held that the city had no right to collect drain water and cast it upon another's land in unnatural volumes and found it liable for damage caused.

The modified common enemy rule resulted in liability in *Oklahoma City v. Bethel*, 175 Okla. 193, 51 P.2d 313 (1935). The city had built a municipal storm sewer system to drain a considerable area of the city. The outlet was to a ditch which was inadequate to carry the collected stormwaters from a heavy rain, causing an overflow and damage to property. The court held that the city had no right to collect water by artificial means and permit it to be discharged in greater volume or velocity than would naturally flow there prior to the construction.

The reasonable use rule was used in an Ohio case. A city had increased the size of an inlet to a ditch in order to accommodate the increased drainage from a new shopping center; however, it failed to increase the size of the outlet from that ditch. Plaintiff's property was between the inlet and the outlet. The appellate court remanded the case to the lower court for a decision in conformity with the reasonable use rule: the city would be liable if it could have protected plaintiff's property from flooding by merely increasing the size of the outlet from the ditch, since that would have been the reasonable thing to do [*Chudzinsky v. City of Sylvania*, 372 N.E.2d 611, 53 Ohio App.2d 151 (1976).

Constructive Taking
A legal theory which can be used only against a governmental entity has been quite successful in drainage cases. It is called *inverse condemnation or constructive taking*. The charge is that the governmental entity, instead of using its powers of condemnation *in advance* for land that it will use, has used the land without payment; therefore, the landowner is now in court seeking reimbursement. In *Pumpelly v. Green Bay Co.*, 80 U.S. 166 (1871), a state dam was built which raised water in Lake Winnebago, Wisconsin, so as to overflow plaintiff's land. The state argued that it had the right to build the dams to improve navigation without legal responsibility for injury. The U.S. Supreme Court held that this amounted to a taking of the land, requiring compensation. Similarly, a California agency built a structure which changed the regime of the river, resulting in flooding of plaintiff's land [*Beckley v. Reclamation Board of State*, 205 Calif.App.2d 734, 23 Calif.Rptr. 428 (1962)]. The board had to buy a permanent flowage easement over plaintiff's land. In *Board of Commissioners of Logan County v. Adler*, 69 Colo. 290, 194 P. 621 (1920), the construction of a county bridge resulted in filling up certain channels of the river. This caused water in time of flood to back up and overflow plaintiff's lands. The county was liable for damages under the state constitutional provision that private property shall not be "taken or damaged" for public use without compensation.

In *Masley v. City of Lorain*, 48 Ohio St. 2d 334, 358 N.E. 2d 596 (1976), the court held that the city's construction of a storm sewer system which discharged more water into a natural creek than the creek could reasonably be expected to handle was a "taking" of the adjacent properties which now regularly were flooded. It should be noted that negligence or nonnegligence are not relevant factors in a decision of liability under the constructive taking theory.

Liability for Granting a Permit to Build

While courts have almost universally held public entities liable for structures or facilities which they have built and which cause damage, there has been a reluctance on the part of some courts to find liability when the public entity merely granted permits to build. In *Miller v. City of Brentwood*, 548 S.W. 2d 878 (Tenn. App. 1977), the court held that the city was not liable for damages due to issuing of building permits, for failure to restrain developers from reducing water absorption of the land, or for failure to provide adequate drainage. The right of action, if any, was against the owners who produced the unnatural drainage.

On the other hand, some courts are finding cities liable in situations where the city issued permits for developments which caused injury to other property owners. In *Elliott v. Nordlof*, 83 Ill. App. 2d 279, 227 N.W. 2d 547 (1967), the new subdivision streets and storm sewers collected surface water which normally did not flow onto plaintiff's property. The court held that the subdivider was not relieved of liability even though the city had accepted the subdivision plat, and the city became liable for the damages when it accepted the plat and the maintenance of the facilities. In *Myotte v. Village of Mayfield*, 54 Ohio App. 2d 97, 375 N.E. 2d 816 (1977), the village had granted a permit to develop an industrial complex which reduced the permeable area and increased flooding. While plaintiff's property had flooded previously, there now was regular flooding every time it rained. The village had told the plaintiff that she should enlarge the culvert on her property to handle the increased flow. But the court held that it was fair and reasonable that the village be liable for granting the permit, awarded damages, and enjoined the village from issuing any more building permits.

These cases recognize that the method of urban development has changed. Instead of a city providing facilities, massive subdivisions are being created by large corporations which are subsequently accepted or annexed as a finished product. They are usually built in conformance with city subdivision or building regulations. Conceptually there is little difference between the city itself collecting and concentrating storm drainage and the city issuing a permit for a development which collects and concentrates storm drainage in accordance with city criteria. In addition, a number of developers may be active in the same watershed. Here the city (or county) is the only entity which sees the total picture and has the overall authority to protect lower property owners. More and more courts are likely to "pierce the corporate veil" of a municipal corporation to find liability in such situations.

In watercourse situations, once again public entities are regularly held liable for flood damage caused by their own construction such as roads, bridges, and dams. However, there appears as yet to be no decision which finds a governmental entity liable for issuing a permit to build a private structure in the floodway or floodplain which diverts or casts more floodwaters onto another's property, causing damage. And yet where the governmental entity knew or should have known that such damage could result, the distinction has little substance. Potential liability in such a situation is suggested in *Cappture Realty Corp. v. Board of Adjustment*, 126 N.J. Sup. 200, 313 A.2d 624 (1973). In upholding a borough's moratorium on building in flood-prone areas the court stated:

> As an interim enactment plaintiff has not been permanently deprived of its property or all uses thereof. In fact, construction which affects the flow of Fleischer Brook so as to damage upstream or down-stream property owners may well subject a landowner who so increases the flow, and possibly a municipality and county, to damage suits. (126 N.J. Sup. at 213, 313 A.2d at 632)

As more of the nation's floodways and floodplains are mapped, local governments will have difficulty in pleading that they were ignorant of the dangers of permitting new structures in such flood hazard areas. Once again, they are the only entities with the overview and overall control of the cumulative effect of a number of such structures. Courts are already recognizing that additional filling and development raise flood heights and alter flows (see Sec. 2.4 under "Protection of Properties Other Than the Regulated Property"). It is a logical next step for a court to find that a local government is liable for permitting such filling or structure which increased the flood damage to other properties when the permitting entity knew, or should have known, that it would. Therefore, adoption of floodplain regulations which control the use of riparian properties to prevent increased flood injury to other riparian properties may be essential to avoid liability under watercourse law.

Governmental Immunity
The theory of governmental immunity as a defense against liability is based on the old English common law principle that "the King can do no wrong." It varies from state to state. In some states, it has been completely repealed [*Ayala v. Philadelphia Board of Public Education*, 453 Pa. 584, 305 A.2d 877 (1973), cited in *Breiner v. C & P Home Builders, Inc.*]. In some states it is still viable. Some states have statutes which specify which governmental entities are immune and to what extent [*Mozzetti v. City of Brisbane*, 67 Calif.App.3d 565, 136 Calif.Rptr. 751 (1977)]. In some cases the distinction is made between the public entity acting in its governmental capacity (not liable) and in its ministerial capacity (liable). The construction and maintenance of drainage facilities are ministerial, so that even in those jurisdictions which recognize governmental immunity, a public body would be liable for its negligence. The design of such facilities, however, is considered a "governmental" function by some courts, so that an inadequate design alone would not cause liability. The distinction is a fine one and was rejected in the *Mozzetti* case where California even had a statute granting "design immunity." The court held that design immunity was not applicable where testimony showed that the city had failed to comply with good engineering practices. It would be foolhardy, therefore, to depend on governmental immunity in the drainage and flood control area.

Use of the Police Power in the Floodplain
It has been indicated that a city may be liable for issuing permits to build and fill floodways and flood hazard areas. Mention must also be made of the potential legal impact when local governments *do* manage floodplain development by zoning, subdivision regulations, and other methods. Such regulations can

be challenged in court on the bases that they violate federal or state constitutional prohibitions against "taking" of private property without compensation, violate equal protection, violate due process, or are beyond the authority vested in the regulating entity. The major cases in this field are discussed in Sec. 2.4; some strike down and others uphold floodplain regulations. Legal counsel is absolutely essential to produce regulations which will result in wise use of the floodplain and be upheld against challenges. If the regulation is struck down, it means that the regulation is unconstitutional as applied to a particular property. The city or county then has the option of permitting the development to occur or purchasing the property at a fair market value. The regulation may also have to be amended to respond to the court's standard. The point is that floodplain regulations which are too strict may at worst be rejected by a court. Floodplain regulations which are too lenient, however, may result in loss of property and lives. In addition, the regulating entity will never know what the true limits of its constitutional authority are.

Conclusions

Potential liability for public and private entities, their engineers, and planners must be factored into the decision-making process for stormwater management. An engineering alternative may lead to additional risk by increasing the development within the flood hazard area. This increased development can create more hazard to life and property than under natural conditions.

2.4 FLOODPLAIN MANAGEMENT

Floodplain management in this section means the exercise of the police power by a governmental entity to control land use in areas subject to inundation. This includes the use of various techniques such as zoning, establishing channel encroachment lines, establishing dune lines, requiring permits to build dams and other obstructions, and requiring railroads to place openings in their embankments for the passage of waters. It also includes controlling dredge and fill and protection of wetlands and tidal marshlands. The cases which have tested these regulations are interrelated philosophically, legally, and historically. Wetlands decisions quote and cite floodplain zoning cases, for example, and vice versa. Some of the traditional zoning cases are also used here as legitimate ancestors to the present generation of cases. Chapter 9 presents additional material on floodplain regulations.

Comparison with Conventional Comprehensive Zoning

Control of land use in general by governmental entities using devices such as zoning is often regarded as governmental interference with private property rights. Yet before zoning was instituted, private parties often had to protect their property rights by going to court. The classic example is a noxious industry moving into a residential area. Under the common law a homeowner would have to file a lawsuit to prevent or abate the "nuisance." With increasing urbanization, a more orderly system had to be worked out. To prevent incompatible uses at the outset, land was segregated into zones for specific uses such as residential, commercial and industrial. Zoning maps were published by cities and counties and became the basis for development.

Such comprehensiv... ...ve land use zoning was upheld by the U.S. Supreme Court
in 1926 in its lan... ...dmark decision, *Village of Euclid* v. *Amber Realty*, 272 U.S.
365 (1926).

...ain zoning also is attacked as governmental interference. Yet
...s have been used to pass floodwaters long before humans appeared
... scene and long before the floodplains became someone's private
...perty. In recent years it has become more evident, as losses to lives
and property have steadily escalated in the United States, that the intensive
human occupancy of such floodplains is incompatible with their natural use of
carrying floodwaters. The major difference between the comprehensive zoning
upheld in *Village of Euclid* and floodplain regulation is that the incompatibility
of uses which are being prevented results not from competing human activities
(for example, industrial versus residential) but from natural origins. Since
Mother Nature's uses are far more permanent and immutable than humankind's,
they should be given the same if not greater consideration by the courts to
prevent future incompatibility. In addition, without floodplain regulation we
still have the situation where private landowners have to sue other private
landowners for damages incurred when a dike or structure built by the other
party causes flooding to their property, instead of preventing such obstructions to the natural flow in the first place. Regulation of the floodplain, then,
protects the rights of private property owners and relieves them of the
burden of protecting themselves from increased damages caused by others.

Judicial Recognition of the Economic Effects of Floodplain
Management

As society becomes more aware that land, water, and air are not commodities
that can be abused without long-term and perhaps even irreversible negative
impacts, state and federal courts are upholding regulations which even a
decade ago would have been struck down. They are broadening their interpretation of public health, safety, and welfare and recognizing an array of
public harms which such regulations are seeking to prevent. Testimony presented in these cases is becoming increasingly scientific, and with this
improved factual basis, courts are making new decisions which will have wide
ranging effects. By floodplain regulations we refer to any array of techniques designed to keep people away from the water, as contrasted to structural measures (dams, dikes, levees, seawalls, etc.) designed to keep the
water away from people. Floodplain regulation techniques include: comprehensive planning; building codes and building permits; floodplain zoning;
... regulations; site plan review; water supply, sewerage, drainage,
and erosion control regulations; utility location regulations; tidal and fresh-
water wetlands regulations; environmental regulations; setback lines; acquisition and regulation. Floodplain regulations are widely accepted as an appropriate exercise of the police power by a duly constituted legislative body.
Regulations are presumed valid if they:

1. Conform to and do not exceed the authority granted in enabling
   statutes (generally related to explicit objectives or implicitly derived
   from health, safety, and welfare provisions)
2. Adhere to the doctrine of reasonableness, i.e., they are not clearly
   arbitrary, having no substantial relationships to public health,
   safety and welfare
3. Forbid discriminatory treatment, i.e., require equal treatment for
   similarly situated properties

A basic tension exists between the rights of the private property owner to
use his property, unencumbered by regulation, and the responsibility of
levels of government for the health, safety, and well-being of their citizens.
As noted in the previous section, landowners bitterly denounce perceived
diminution in property values resulting from floodplain regulations. Governments are accused of taking property without just compensation; rendering
unmarketable private property located in identified floodways; forcing individuals to sell property or to pay taxes not commensurate with the use permitted; and forcing individuals to incur increased construction costs that
result from such regulations.

Change in Market Value and Economic Uses of the Property Being Regulated

The most common attack on land use controls is that regulations are an unconstitutional taking of private property without just compensation, in violation of the Fifth Amendment to the Constitution. In *Pennsylvania Coal Co.
v. Mahon*, 260 U.S. 393 (1922), the U.S. Supreme Court ruled that one of the
considerations in deciding whether a regulation exceeds its constitutional
limits is the degree to which the value of the property has been diminished.
In *Village of Euclid v. Ambler* the Court was not persuaded by a value
differential, holding that before a zoning regulation can be declared
unconstitutional it must be found to be clearly arbitrary and unreasonable,
having no substantial relation to public health, safety, morals, or general
welfare. Then in 1962 the Court reiterated that there is no set formula
to determine where legitimate regulation ends and taking begins, and
that while a comparison of values before and after regulation is relevant
(citing *Pennsylvania Coal*), it is by no means conclusive [*Goldblatt v.
Hempstead*, 369 U.S. 590 (1962)]. State courts have universally maintained
that the stringency of regulations must be reasonably related to the severity
of the public harm being mitigated by the regulations [*MacGibbon v. Board
of Appeals of Duxbury*, 356 Mass. 696, 255 N.E.2d 347 (1970)]. After a number of cases across the country found taking to be the issue, in Massachusetts
the court held that even a substantial (88 percent) decrease in value of rezoned property is not a conclusive argument against the rezoning [*Turnpike
Realty v. Town of Dedham*, 362 Mass. 221, 284 N.E.2d 891 (1972), certiorari
denied 409 U.S. 1108 (1973)].

With little fanfare state courts in Wisconsin and New Hampshire have
taken a giant step on the issue of the value of the property being regulated.
Prior courts had almost always focused on the market value, on the
"value" of the property as viewed in the real estate market in terms of the
profit which can be derived from the land for an individual or small segment
of society. These two courts, however, have analyzed the worth of the
property in light of what might be called its *intrinsic* value, based on its role
in the ecosystem, related to its value to society as a whole.

In *Just v. Marinette County*, 56 Wis.2d 7, 201 N.W.2d 761 (1972), the
court held that an owner has no absolute and unlimited right to change the
essential natural character of his land so as to use it for a purpose for which
it is unsuited in its natural state and which injures the rights of others
(here, the rights of the public to preserve the natural environment and the
natural relationship between the wetlands and the purity of the water and
natural resources). Regarding value, the court pointed out that the alleged

reciation was not based on the use of the land in its natural state, but on
t land would be worth if it could be filled.

While loss of value is to be considered in determining whether a restriction
is a constructive taking, value based upon *changing* the character of the
land is not an essential factor or controlling. (201 N.W. 2d at 771)

The New Hampshire court embraced this rationale in *Sibson v. State*,
 N.H. 124, 336 A.2d 239 (1975), stressing:

The denial of the permit did not depreciate the value of the marshland or
cause it to become "of practically no pecuniary value." Its value was
the same after the denial of the permit as before and it remained as it
had been for millenniums. (115 N.H. at 129, 336 A.2d at 243)

e court further stated that no taking had occurred because the denial pre-
nted public harm (rather than created a public benefit). Under the old
nnsylvania Coal diminution of value test, however, another logical conclu-
on of the argument above is that since there is no change in value, there is
viously no diminution in value in such circumstances, and therefore there
 also no taking.

The most important role of floodplain land may be that of carrying flood-
aters, especially since the filling or development of such land shifts the
rden of carr loodwaters onto other lands which in the original, natural
      d. o   t have to function in that role at all. In addition, the
   h is he   lue of property in a hazard area was unrecognized or
   a shift to   may be on a geologic fault, an old mudslide, or
   in a non-   n these geologic hazards become widely known, its
   omic dis-   s price should decrease, with or without regula-
                 nply be a recognition of this unsuitability by the

   in cases   r Than the Regulated
   age to pri-
   Commission.   properties other than regulated property. A
   State].   e hazards of floodplain encroachment on other
   ble for   nty of Del Norte, 24 Calif.App.3d 311, 101
   s to   urt stated that the floodplain regulations which
   r,   a subdivision subject to flooding, and permitted
   oodplain.   lture, were a valid exercise of the police
   ccurence of a   buildings would increase flood heights
   d the age-old   er buildings outside the zoned area.
   more of the   r a temporary moratorium [*Cappture*
   ts will have-   rejection of fill permits [*Turnpike*
   new develop-   ial of building permits [*Vazza*
   n regard to   n, 296 N.E.2d 220 (Mass.Ct.App.
   ecognizing that   affect flood stages upstream,
   flows, and in-
   the next step is
   mitting such   nce with
   rds.
                           one where, in order to develop, the
                           or fill placed on the site to raise it
                           ial costs to comply with floodplain

regulations are usually involved. While there are no cases as yet specificall
testing such building regulations, conceptually they are no different than
others which local governments have established in their building codes or
subdivision regulations. The necessary expense or loss of value which will
be sustained by the property owner as a result of the regulations does not
invalidate such regulations, e.g., *City of Chicago v. Washington Home of
Chicago*, 289 Ill. 206, 124 N.E. 416 (1919).

Loss of Tax Base

One argument raised in opposition to floodplain/wetland regulation is that
otherwise developable land cannot be developed, causing a hardship on the
community because of a loss of tax base. No case has been found to suppor
this view. Instead, courts have enumerated public benefits gained and
public harm prevented. Even those courts which have struck down regulations have recognized the benefits, but have decided that to obtain such
benefits, eminent domain rather than police power was the legitimate tool
[*State v. Johnson*, 265 A.2d 711 (Me. 1970); *MacGibbon v. Board of Appeals
of Duxbury; Morris County Land Improvement Company v. Parsippany-Troy
Hill Township*, 40 N.J. 539, 193 A.2d. 232 (1963)].

There is recognition, on the other hand, that a regulation may protect
the tax base [see *Just v. Marinette County; Cappture Realty Corp. v.
Board of Adjustment*]. Even assuming that regulations do result in a
lower assessed valuation for floodplain properties, there may still not b
tax base loss in totality for the community. Often there is other la
side the flood hazard area but still within the tax jurisdiction, whic
suited for the development. In that case there is no net loss, just
a more appropriate location. In addition, locating the development
hazardous location could prevent a future loss of tax base and econ
ruption as a result of a flood.

Avoidance of Public Liability

The finding of public liability in the floodplain has generally been
where a governmental entity has built something which caused dam
vate property in times of flooding [*Barr v. Game, Fish and Parks
City of Vicksburg v. Porterfield, Beckley v. Reclamation Board of*

There is as yet no decision that finds a government entity lia
allowing a private structure to be built which causes flood damage
another's property. Such potential liability is suggested, howeve
Cappture Realty.

There is an increasing potential for public liability in the fl
With improved meteorological and hydrological information, the
flood of a certain magnitude becomes a statistical probability, a
rationale for negligence as an act of God becomes suspect. As
nation's floodways and floodplains are mapped, local governme
difficulty pleading that they are ignorant of the dangers whe
ments, which they permitted, are destroyed and lives lost.
existing development in the floodplains, courts are already
additional filling and development raise flood heights, alter
crease the hazard to the existing development. Logically,
likely to be the finding of a local government liable for pe
development when it had knowledge of potential flood haz

Judicial Recognition of the Social Effects of Floodplain
Management

Social effects recognized by the courts relate to protecting the health, safety,
and welfare of floodplain residents.

Developers of floodplains, who may or may not be aware of the risk, are
in and out of the property within a few years, selling it to unwary buyers.
Floodplain encroachment affects other individuals and the public. The loss of
lives and property, of public facilities such as utility lines and roads to serve
floodplain development, or of employment centers, closed down because of
flood damage, has a detrimental and long-term effect on the community as a
whole.

Protection of Lives and Property, Including the Urban
Infrastructure

Preservation of public health, safety, and welfare is the cornerstone
for the exercise of police powers.

Regulations to minimize threats to public safety enjoy a presumption of
constitutionality [*Biffer v. City of Chicago*, 278 Ill. 562, 116 N.E. 182
(1917)]. The *Turnpike Realty* case recognized that restrictions on land serve
to protect those who might choose to develop or occupy the land in spite of
the dangers to themselves or their property. In *Turner v. County of Del
Norte*, the court held that where evidence showed a frequency of flooding
which would almost certainly eventually destroy permanent residences and
endanger the lives and health of their occupants, such developments can be
prevented by zoning (see also *Vazza Properties*).

In *Spiegle v. Borough of Beach Haven*, 116 N.J. Super. Ct. 148, 281 A.2d
377 (1971), the court recognized both the need to protect the owner from his
own folly and the need to protect the public from his indiscretion. The court
upheld the Borough's prohibition of building seaward of the dune line of the
beachfront where construction would be subjected to hazards from wind and
water. The court stressed the fact that ruptured sewer lines could endanger
the borough's entire sewer system, a ruptured water line might result in the
municipal tank being drained, and a ruptured gas line would also create a
dangerous condition. Road service could not be provided or feasibly maintained over the beach.

Preclusion of Need for Public Expenditure for Protective Works
and Disaster Relief

As more and more development occurs in the floodplain, the flooding situation
is aggravated, and there is clamor to build protective works such as dams,
dikes, and channels. At least one court has recognized that expending public
dollars for such protective works may actually be a subsidy to private
development. In *Cappture Realty* the court upheld a building moratorium
until a flood control project could be built, and stated:

> Although perhaps unnecessary for this decision, the other side of the
> coin is often overlooked in analyzing claims by a landowner in a flood
> area that he should be able to use his land without restriction. If pri-
> vate construction would call for, or perhaps demand or increase the de-
> mand for, public flood control projects, does this not call for an expendi-
> ture of public funds for the protection of a specific private property,
> or purposes?

The *Turnpike Realty* court upheld the constitutionality of floodplain regulations which have as a major objective the protection of the entire community from individual choices of land use which require subsequent public expenditures for public works and disaster relief.

Judicial Recognition of the Environmental Effects of Floodplain Management

Courts have recognized the natural functions of the floodplain and the prob-lems that arise as a result of their loss. The reduction of storage capacity, groundwater recharge area, water quality, ecosystem quality, and recreation-al areas has economic and social dimensions that transcend the land that is subject to regulation.

Preservation of the Flood-Carrying Capacity and Storage Capacity

A U.S. Supreme Court decision, *Chicago & Alton R. R. v. Tranbarger*, 238 U.S. 67 (1915), held that an act to prevent railroad embankments from de-flecting surface water from its usual course, thereby injuring the land of another, was a legitimate regulation established under the state's police power.

In *City of Welch v. Mitchell*, 95 W.Va. 377, 121 S.E. 165 (1924), the court held that the city has the right to set equitable building lines on either side of a creek equidistant from the center in order to prevent obstruction of the flow of the stream.

In *Vartelas v. Water Resources Commission*, 146 Conn. 650, 153 A.2d 822 (1959), the court found the establishment of encroachment lines a valid means of maintaining the capacity of a channel and avoiding raising flood stages. *Iowa Natural Resources Council v. Van Zee*, 261 Iowa 1287, 158 N.W.2d 111 (1968) also recognizes the importance of this issue.

The *Morris County Land Improvement Company* court recognized the value of preserving swampland as a natural stormwater detention basin for waters, but stated that its preservation could not be accomplished by an exercise of police power. To obtain the public benefit, compensation for the land would have to be made. The *Vazza Properties* court, however, came to a different conclusion, holding that building a large apartment complex and parking area would aggravate a periodic flooding problem in nearby residential areas by eliminating a natural soft-peat holding area.

Preservation of Groundwater Recharge Area

Cases in the area of water law have long recognized the hydraulic connection between surface waters and groundwaters, especially in the semiarid West. In the California case of *Miller v. Bay Cities Water Co.*, 157 Calif. 256, 107 P. 115 (1910), the court permanently enjoined a water supply corporation from building a dam, finding that the plaintiff, an orchard owner, had a right to the continued flows of floodwaters to recharge his aquifer; these flood-waters were those which could reasonably be anticipated during ordinary seasons.

In *Turnpike Realty*, the court upheld the town's ordinance establishing a floodplain district as a valid means of preserving and maintaining the groundwater table.

Such comprehensive land use zoning was upheld by the U.S. Supreme Court
in 1926 in its landmark decision, *Village of Euclid v. Amber Realty*, 272 U.S.
365 (1926).

Floodplain zoning also is attacked as governmental interference. Yet
floodplains have been used to pass floodwaters long before humans appeared
on the scene and long before the floodplains became someone's private
property. In recent years it has become more evident, as losses to lives
and property have steadily escalated in the United States, that the intensive
human occupancy of such floodplains is incompatible with their natural use of
carrying floodwaters. The major difference between the comprehensive zoning
upheld in *Village of Euclid* and floodplain regulation is that the incompatibility
of uses which are being prevented results not from competing human activities
(for example, industrial versus residential) but from natural origins. Since
Mother Nature's uses are far more permanent and immutable than humankind's,
they should be given the same if not greater consideration by the courts to
prevent future incompatibility. In addition, without floodplain regulation we
still have the situation where private landowners have to sue other private
landowners for damages incurred when a dike or structure built by the other
party causes flooding to their property, instead of preventing such obstructions to the natural flow in the first place. Regulation of the floodplain, then,
protects the rights of private property owners and relieves them of the
burden of protecting themselves from increased damages caused by others.

Judicial Recognition of the Economic Effects of Floodplain
Management

As society becomes more aware that land, water, and air are not commodities
that can be abused without long-term and perhaps even irreversible negative
impacts, state and federal courts are upholding regulations which even a
decade ago would have been struck down. They are broadening their interpretation of public health, safety, and welfare and recognizing an array of
public harms which such regulations are seeking to prevent. Testimony presented in these cases is becoming increasingly scientific, and with this
improved factual basis, courts are making new decisions which will have wide-
ranging effects. By floodplain regulations we refer to any array of techniques designed to keep people away from the water, as contrasted to structural measures (dams, dikes, levees, seawalls, etc.) designed to keep the
water away from people. Floodplain regulation techniques include: compre
hensive planning; building codes and building permits; floodplain zoning;
subdivision regulations; site plan review; water supply, sewerage, drainage,
and erosion control regulations; utility location regulations; tidal and fresh-
water wetlands regulations; environmental regulations; setback lines; acquisition and regulation. Floodplain regulations are widely accepted as an appropriate exercise of the police power by a duly constituted legislative body.
Regulations are presumed valid if they:

1. Conform to and do not exceed the authority granted in enabling
   statutes (generally related to explicit objectives or implicitly derived
   from health, safety, and welfare provisions)
2. Adhere to the doctrine of reasonableness, i.e., they are not clearly
   arbitrary, having no substantial relationships to public health,
   safety and welfare
3. Forbid discriminatory treatment, i.e., require equal treatment for
   similarly situated properties

A basic tension exists between the rights of the private property owner to
use his property, unencumbered by regulation, and the responsibility of all
levels of government for the health, safety, and well-being of their citizens.
As noted in the previous section, landowners bitterly denounce perceived
diminution in property values resulting from floodplain regulations. Governments are accused of taking property without just compensation; rendering
unmarketable private property located in identified floodways; forcing individuals to sell property or to pay taxes not commensurate with the use permitted; and forcing individuals to incur increased construction costs that
result from such regulations.

Change in Market Value and Economic Uses of the Property Being Regulated

The most common attack on land use controls is that regulations are an unconstitutional taking of private property without just compensation, in violation of the Fifth Amendment to the Constitution. In *Pennsylvania Coal Co.
v. Mahon,* 260 U.S. 393 (1922), the U.S. Supreme Court ruled that one of the
considerations in deciding whether a regulation exceeds its constitutional
limits is the degree to which the value of the property has been diminished.
In *Village of Euclid v. Ambler* the Court was not persuaded by a value
differential, holding that before a zoning regulation can be declared
unconstitutional it must be found to be clearly arbitrary and unreasonable,
having no substantial relation to public health, safety, morals, or general
welfare. Then in 1962 the Court reiterated that there is no set formula
to determine where legitimate regulation ends and taking begins, and
that while a comparison of values before and after regulation is relevant
(citing *Pennsylvania Coal*), it is by no means conclusive [*Goldblatt v.
Hempstead,* 369 U.S. 590 (1962)]. State courts have universally maintained
that the stringency of regulations must be reasonably related to the severity
of the public harm being mitigated by the regulations [*MacGibbon v. Board
of Appeals of Duxbury,* 356 Mass. 696, 255 N.E.2d 347 (1970)]. After a number of cases across the country found taking to be the issue, in Massachusetts
the court held that even a substantial (88 percent) decrease in value of rezoned property is not a conclusive argument against the rezoning [*Turnpike
Realty v. Town of Dedham,* 362 Mass. 221, 284 N.E.2d 891 (1972), certiorari
denied 409 U.S. 1108 (1973)].

     With little fanfare state courts in Wisconsin and New Hampshire have
taken a giant step on the issue of the value of the property being regulated.
Prior courts had almost always focused on the *market* value, that is, the
"value" of the property as viewed in the real estate market in terms of the
profit which can be derived from the land for an individual or small segment
of society. These two courts, however, have analyzed the worth of the
property in light of what might be called its *intrinsic* value, based on its role
in the ecosystem, related to its value to society as a whole.

     In *Just v. Marinette County,* 56 Wis.2d 7, 201 N.W.2d 761 (1972), the
court held that an owner has no absolute and unlimited right to change the
essential natural character of his land so as to use it for a purpose for which
it is unsuited in its natural state and which injures the rights of others
(here, the rights of the public to preserve the natural environment and the
natural relationship between the wetlands and the purity of the water and
natural resources). Regarding value, the court pointed out that the alleged

depreciation was not based on the use of the land in its natural state, but on what land would be worth if it could be filled.

> While loss of value is to be considered in determining whether a restriction is a constructive taking, value based upon *changing* the character of the land is not an essential factor or controlling. (201 N.W. 2d at 771)

The New Hampshire court embraced this rationale in *Sibson v. State*, 115 N.H. 124, 336 A.2d 239 (1975), stressing:

> The denial of the permit did not depreciate the value of the marshland or cause it to become "of practically no pecuniary value." Its value was the same after the denial of the permit as before and it remained as it had been for millenniums. (115 N.H. at 129, 336 A.2d at 243)

The court further stated that no taking had occurred because the denial prevented public harm (rather than created a public benefit). Under the old *Pennsylvania Coal* diminution of value test, however, another logical conclusion of the argument above is that since there is no change in value, there is obviously no diminution in value in such circumstances, and therefore there is also no taking.

The most important role of floodplain land may be that of carrying floodwaters, especially since the filling or development of such land shifts the burden of carrying floodwaters onto other lands which in the original, natural order of things did not have to function in that role at all. In addition, the preregulation *market* value of property in a hazard area was unrecognized or not widely known. Land may be on a geologic fault, an old mudslide, or subject to avalanches. When these geologic hazards become widely known, its development potential and its price should decrease, with or without regulation. The regulation may simply be a recognition of this unsuitability by the general public.

Protection of Properties Other Than the Regulated
Property
Floodplain controls also affect properties other than regulated property. A few courts have recognized the hazards of floodplain encroachment on other properties. In *Turner v. County of Del Norte*, 24 Calif.App.3d 311, 101 Calif. Rptr. 93 (1972), the court stated that the floodplain regulations which prohibited further building in a subdivision subject to flooding, and permitted only parks, recreation, and agriculture, were a valid exercise of the police power. Evidence showed that such buildings would increase flood heights which could increase the hazard to other buildings outside the zoned area. Other similar examples include imposing a temporary moratorium [*Cappture Realty Corp. v. Board of Adjustment*], rejection of fill permits [*Turnpike Realty Co. v. Town of Dedham*], and denial of building permits [*Vazza Properties, Inc. v. City Council of Woburn*, 296 N.E.2d 220 (Mass.Ct.App. 1973)], where proposed construction would affect flood stages upstream, downstream, or on adjacent properties.

Higher Cost of Development Due to Compliance with
Regulations
If the property is located in a floodplain zone where, in order to develop, the structure would have to be flood-proofed or fill placed on the site to raise it above the regulatory flood level, additional costs to comply with floodplain

regulations are usually involved. While there are no cases as yet specifically
testing such building regulations, conceptually they are no different than
others which local governments have established in their building codes or
subdivision regulations. The necessary expense or loss of value which will
be sustained by the property owner as a result of the regulations does not
invalidate such regulations, e.g., *City of Chicago v. Washington Home of
Chicago*, 289 Ill. 206, 124 N.E. 416 (1919).

Loss of Tax Base

One argument raised in opposition to floodplain/wetland regulation is that
otherwise developable land cannot be developed, causing a hardship on the
community because of a loss of tax base. No case has been found to support
this view. Instead, courts have enumerated public benefits gained and
public harm prevented. Even those courts which have struck down regulations have recognized the benefits, but have decided that to obtain such
benefits, eminent domain rather than police power was the legitimate tool
[*State v. Johnson*, 265 A.2d 711 (Me. 1970); *MacGibbon v. Board of Appeals
of Duxbury; Morris County Land Improvement Company v. Parsippany-Troy
Hill Township*, 40 N.J. 539, 193 A.2d. 232 (1963)].

There is recognition, on the other hand, that a regulation may protect
the tax base [see *Just v. Marinette County; Cappture Realty Corp. v.
Board of Adjustment*]. Even assuming that regulations do result in a
lower assessed valuation for floodplain properties, there may still not be a
tax base loss in totality for the community. Often there is other land, outside the flood hazard area but still within the tax jurisdiction, which is better
suited for the development. In that case there is no net loss, just a shift to
a more appropriate location. In addition, locating the development in a non-
hazardous location could prevent a future loss of tax base and economic disruption as a result of a flood.

Avoidance of Public Liability

The finding of public liability in the floodplain has generally been in cases
where a governmental entity has built something which caused damage to private property in times of flooding [*Barr v. Game, Fish and Parks Commission,
City of Vicksburg v. Porterfield, Beckley v. Reclamation Board of State*].

There is as yet no decision that finds a government entity liable for
allowing a private structure to be built which causes flood damages to
another's property. Such potential liability is suggested, however, in
Cappture Realty.

There is an increasing potential for public liability in the floodplain.
With improved meteorological and hydrological information, the occurence of a
flood of a certain magnitude becomes a statistical probability, and the age-old
rationale for negligence as an act of God becomes suspect. As more of the
nation's floodways and floodplains are mapped, local governments will have
difficulty pleading that they are ignorant of the dangers when new developments, which they permitted, are destroyed and lives lost. In regard to
existing development in the floodplains, courts are already recognizing that
additional filling and development raise flood heights, alter flows, and increase the hazard to the existing development. Logically, the next step is
likely to be the finding of a local government liable for permitting such
development when it had knowledge of potential flood hazards.

Judicial Recognition of the Social Effects of Floodplain
Management

Social effects recognized by the courts relate to protecting the health, safety, and welfare of floodplain residents.

Developers of floodplains, who may or may not be aware of the risk, are in and out of the property within a few years, selling it to unwary buyers. Floodplain encroachment affects other individuals and the public. The loss of lives and property, of public facilities such as utility lines and roads to serve floodplain development, or of employment centers, closed down because of flood damage, has a detrimental and long-term effect on the community as a whole.

*Protection of Lives and Property, Including the Urban
Infrastructure*

Preservation of public health, safety, and welfare is the cornerstone for the exercise of police powers.

Regulations to minimize threats to public safety enjoy a presumption of constitutionality [*Biffer v. City of Chicago*, 278 Ill. 562, 116 N.E. 182 (1917)]. The *Turnpike Realty* case recognized that restrictions on land serve to protect those who might choose to develop or occupy the land in spite of the dangers to themselves or their property. In *Turner v. County of Del Norte*, the court held that where evidence showed a frequency of flooding which would almost certainly eventually destroy permanent residences and endanger the lives and health of their occupants, such developments can be prevented by zoning (see also *Vazza Properties*).

In *Spiegle v. Borough of Beach Haven*, 116 N.J. Super. Ct. 148, 281 A.2d 377 (1971), the court recognized both the need to protect the owner from his own folly and the need to protect the public from his indiscretion. The court upheld the Borough's prohibition of building seaward of the dune line of the beachfront where construction would be subjected to hazards from wind and water. The court stressed the fact that ruptured sewer lines could endanger the borough's entire sewer system, a ruptured water line might result in the municipal tank being drained, and a ruptured gas line would also create a dangerous condition. Road service could not be provided or feasibly maintained over the beach.

*Preclusion of Need for Public Expenditure for Protective Works
and Disaster Relief*

As more and more development occurs in the floodplain, the flooding situation is aggravated, and there is clamor to build protective works such as dams, dikes, and channels. At least one court has recognized that expending public dollars for such protective works may actually be a subsidy to private development. In *Cappture Realty* the court upheld a building moratorium until a flood control project could be built, and stated:

> Although perhaps unnecessary for this decision, the other side of the coin is often overlooked in analyzing claims by a landowner in a flood area that he should be able to use his land without restriction. If private construction would call for, or perhaps demand or increase the demand for, public flood control projects, does this not call for an expenditure of public funds for the protection of a specific private property, or purposes?

The *Turnpike Realty* court upheld the constitutionality of floodplain regulations which have as a major objective the protection of the entire community from individual choices of land use which require subsequent public expenditures for public works and disaster relief.

Judicial Recognition of the Environmental Effects of Floodplain Management

Courts have recognized the natural functions of the floodplain and the problems that arise as a result of their loss. The reduction of storage capacity, groundwater recharge area, water quality, ecosystem quality, and recreational areas has economic and social dimensions that transcend the land that is subject to regulation.

Preservation of the Flood-Carrying Capacity and Storage Capacity

A U.S. Supreme Court decision, *Chicago & Alton R. R. v. Tranbarger*, 238 U.S. 67 (1915), held that an act to prevent railroad embankments from deflecting surface water from its usual course, thereby injuring the land of another, was a legitimate regulation established under the state's police power.

In *City of Welch v. Mitchell*, 95 W.Va. 377, 121 S.E. 165 (1924), the court held that the city has the right to set equitable building lines on either side of a creek equidistant from the center in order to prevent obstruction of the flow of the stream.

In *Vartelas v. Water Resources Commission*, 146 Conn. 650, 153 A.2d 822 (1959), the court found the establishment of encroachment lines a valid means of maintaining the capacity of a channel and avoiding raising flood stages. *Iowa Natural Resources Council v. Van Zee*, 261 Iowa 1287, 158 N.W.2d 111 (1968) also recognizes the importance of this issue.

The *Morris County Land Improvement Company* court recognized the value of preserving swampland as a natural stormwater detention basin for waters, but stated that its preservation could not be accomplished by an exercise of police power. To obtain the public benefit, compensation for the land would have to be made. The *Vazza Properties* court, however, came to a different conclusion, holding that building a large apartment complex and parking area would aggravate a periodic flooding problem in nearby residential areas by eliminating a natural soft-peat holding area.

Preservation of Groundwater Recharge Area

Cases in the area of water law have long recognized the hydraulic connection between surface waters and groundwaters, especially in the semiarid West. In the California case of *Miller v. Bay Cities Water Co.*, 157 Calif. 256, 107 P. 115 (1910), the court permanently enjoined a water supply corporation from building a dam, finding that the plaintiff, an orchard owner, had a right to the continued flows of floodwaters to recharge his aquifer; these floodwaters were those which could reasonably be anticipated during ordinary seasons.

In *Turnpike Realty*, the court upheld the town's ordinance establishing a floodplain district as a valid means of preserving and maintaining the groundwater table.

Protection of Water Quality
The need to protect the quality of water for beneficial uses was recognized
by the U.S. Supreme Court as early as 1931 in *New Jersey v. New York*, 283
U.S. 336, in which Mr. Justice Holmes made his oft-quoted statement, "A
river is more than an amenity, it is a treasure."

A major effect of building in floodplains and wetlands is the degradation
of water quality and aquatic life due to increased pollution and siltation of the
waters. In *Zabel v. Tabb*, 430 F.2d 199 (5th Cir. 1970), certiorari denied 401
U.S. 910 (1971), the court ruled that not only did the federal government have
the power to prohibit for ecological reasons dredging and filling in private
riparian lands submerged in navigable waters, but that the Army Corps of
Engineers was compelled to take such factors into account.

To similar effect, the district court in *United States v. Lewis*, 355
F.Supp. 1132 (S.D.Ga. 1973) pointed out that it was necessary to maintain
the unrestricted ebb and flow through the network of small tidal streams of
the saltwater marsh because this had a cleansing effect on plant and animal
life in the marshland.

This principle is upheld in state courts. In *Candlestick Prop., Inc. v.
San Francisco Bay C. & D. Com'n.*, 11 Calif.App.3d 557, 89 Calif.Rptr. 897
(1970), the court upheld the denial of a fill application, citing the state act
which stated that further piecemeal filling of the bay may adversely affect the
quality of the bay waters and even the quality of air in the bay area.

The *Just* court based its decision in great part on the prevention of
future pollution and eradication of present pollution. Recognizing the effect
of the wetlands and the natural environment of shorelands on the purity of
the water, this court reaffirmed that laws to prevent pollution and protect
the waters from degradation were valid police power enactments.

Two years later, in *State v. Deetz*, 66 Wis.2d 1, 224 N.W.2d 407 (1974),
the same court overturned the traditional "common enemy" rule whereby a
landowner has an unrestricted right to deal with surface water on his land as
he pleases, regardless of the harm which he may cause to others. The court
overturned this precedent and adopted the "reasonable use" doctrine of surface waters and remanded the case to the lower court to determine whether
the conduct of the developer was reasonable.

In *Brecciaroli v. Connecticut Com'r. of Env. Protec.* 168 Conn. 349,
362 A.2d 948 (1975), the court, ruling against a dredge and fill operation,
stated that the "evils" of unreasonable pollution, impairment or destruction of
the state's natural resources, were proper subjects for police power regulations. [See also *Potomac Sand & Gravel Co. v. Governor of Maryland*, 266
Md. 358, 293 A.2d 241 (1972), certiorari denied, 409 U.S. 1040.]

*Preservation of the Environment: Ecosystems, Natural
Resources, Habitat, Fish, and the Production of Nutrients*
There was a major turnaround in the 1970s by the courts on behalf of the
environment. The traditional public health, safety, and welfare focus on
humans alone has been broadened to include the role of humans in the balance
of nature. The federal cases which recognize not only the right, but the
duty, of the Army Corps of Engineers to consider ecological factors in making
decisions on dredge and fill permits are examples of this concern for the
environment.

In *United States v. Lewis*, 355 F.Supp. 1132 (S.D.Ga. 1973), the court
pointed to the importance of the productivity of the marshlands as a primary

energy source, providing a basic unit in the food chain of sea animal life. It quoted a report which showed that a Georgia salt marsh can outperform an average wheat field severalfold in organic production.

In *Rivers Defense Committee v. Thierman*, 380 F.Supp. 91 (S.D. N.Y. 1974), the court found that irreparable injury could occur to spawning and nursery areas for many fish species as a result of fill [see also *Zabel v. Tabb*, and *United States v. Joseph G. Moretti*, 331 F.Supp. 151 (S.D.Fla. 1971)]. In *Sands Point Harbor, Inc. v. Sullivan*, 136 N.J.Super. 436, 346 A.2d 612 (1975), the New Jersey court held that regulation of the use of marshes and wetlands which have environmental and ecological importance to the continued existence of species and to humankind is a valid exercise of governmental power.

Providing for Recreation
The mention of public recreational benefits incidental to legitimate purposes has been avoided in statutes regulating flood-prone areas because such regulation is a use of police power. However, in *Turnpike Realty* the purposes clause of the regulation listed the preservation of open space for education and recreation. The court specifically states that while aesthetic considerations alone would not justify the regulation, it was fully supported by the other purposes listed, and the open space purpose would not invalidate the regulation.

In federal dredge and fill cases, however, the federal authority to regulate stems from the Commerce Clause of the U.S. Constitution rather than police power. Therefore, recreational benefits can be mentioned. In *Zabel* the court quoted extensively from various acts and the U.S. Army Corps of Engineers' own implementing regulations which require consideration of recreational aspects when making decisions on dredge and fill permits.

Conclusions

The most common constitutional attack on state (or local) laws is the prohibition against taking without just compensation. In upholding such laws, courts couch their decisions in terms of public harms which are being prevented rather than public benefits which are being obtained. Almost any action can be viewed from different perspectives. For example, the prevention of wetland development can be seen as preventing the loss of a natural resource which would be a public harm, or it can be viewed as an effort to gain a flood retention pond for the benefit of the general public. Federal laws, on the other hand, are vulnerable to the attack that they are beyond the authority granted to Congress in the U.S. Constitution. Once federal authority is upheld, such as in the U.S. Army Corps of Engineers dredge and fill cases, the courts can cite the public benefits which the law sought to attain. State courts (and federal courts when acting on state and local regulations) cite harm to be prevented, while cases based on federal laws can cite benefits to be obtained. In either situation, there is no doubt that the present overall judicial thrust in the United States is to uphold floodplain regulations based on an ever-expanding array of economic, social, and environmental justifications.

2.5 WATER POLLUTION CONTROL

Engineering studies have shown that a substantial amount of pollution is carried into the nation's waters by stormwater runoff. In urban areas such pollution typically includes sediment, oil, grease, salts, and human and animal debris. Construction activities, leaching from mine tailings, and pesticides and fertilizers from agricultural lands are also sources of pollution. The stormwater carries these pollutants into streams and lakes by overland flow or discharges them from storm sewers. Overall degradation of the water quality occurs, and the initial shock loading of pollutants may be particularly detrimental to aquatic life.

Federal Act

Present federal policy in water pollution control is articulated in the Clean Water Act (PL 92-500 as amended by PL 95-217, 33 U.S.C. sections 1251 et seq.). Its objective is to restore and maintain the chemical, physical, and biological integrity of the nation's water. It establishes an interim goal that, wherever attainable, water quality which provides for the protection and propagation of fish, shellfish, and wildlife and provides for recreation in and on the water be achieved by 1983. The ultimate 1985 goal is to eliminate the discharge of pollutants into the nation's water. Many state water pollution control acts have similar goals and objectives.

The federal act recognizes that pollution is caused not only by discharges from municipal and industrial sources, but by stormwater runoff. The act speaks in terms of point sources and nonpoint sources of pollution. A *point source* is defined as any discernible, confined, and discrete conveyance, including but not limited to any pipe, ditch, channel, tunnel, conduit, or similar conveyance, from which pollutants are discharged. *Nonpoint sources*, however, are diffuse sources such as overland runoff, construction activities, and mine drainage. Urban stormwater runoff can be either point or nonpoint, depending on whether it enters the stream from a storm sewer or from diffused overland flow. The distinction is important because the control techniques established by the act differ depending on whether the source is point or nonpoint.

Control of Storm Sewers (Point Source Pollution)

Point sources are controlled by National Pollution Discharge Elimination System (NPDES) permits. Initially the Environmental Protection Agency (EPA) exempted from the NPDES program those storm sewers which were composed entirely of storm runoff uncontaminated by commercial or industrial activity. (It should be pointed out, however, that storm sewers which either the EPA or a state water pollution agency identified as a "significant contributor of pollution" were never exempted—38 FR 18000, July 5, 1973.) This exemption, among others, was challenged in *Natural Resources Defense Counsel, Inc. v. Train*, 396 F.Supp. 1393 (D.C.Cir. 1975). The court held that the EPA did not have the latitude to exempt such point sources. The decision was upheld in *Natural Resources Defense Counsel, Inc. v. Costle*, 568 F.2d 1369 (1977). EPA regulations promulgated in response to the district court's order (41 FR 11303, March 18, 1976, and 42 FR 6846, February 4, 1977) have recently been revised and incorporated into a full set of revised comprehensive regulations for the NPDES program (44 FR 32853, June 7, 1979; see in particular sections 122.45 and 122.48). The permit system for storm sewers is briefly summarized as follows:

1. Storm sewers which discharge water contaminated by contact with wastes, raw materials, or contaminated soil from industrial or commerical facilities into waters of the United States, or into "separate storm sewers,"

are subject to the individual NPDES permit program. Note that an NPDES permit can be issued for discharge *into* a storm sewer.

2. A *separate storm sewer* means a conveyance primarily used for conveying stormwater runoff and located either (a) in an urbanized area, or (b) not in an urbanized area but designated as a "significant contributor of pollution." An urbanized area is essentially a population center of 50,000 or more, located within a Standard Metropolitan Statistical Area (SMSA), on the basis of the 1980 census. The designation on a case-by-case basis in the nonurban area considers such factors as size, location, quantity, and nature of the pollutants. Designation can also be made when approving a section 208 areawide water quality management plan which contains requirements applicable to storm sewers.

3. A *general permit* can be issued to all owners or operators of separate storm sewers in a designated "general permit program area" to authorize discharge into water of the United States if, among other factors, they (a) discharge the same type of waste; (b) would require the same effluent limitations or operating conditions; and (c) would require similar monitoring requirements.

4. A general permit issued to any person may be revoked and an individual NPDES permit required. Such revocation may be based on various factors including the facts that the discharge is a significant contributor of pollution, the discharge is not complying with the general permit, or a change has occurred in the technology to control or abate the pollutants.

It is important to note the words *effluent limitations or operating conditions* in 3(b) above. An argument by the defendant EPA in cases cited above was that requiring NPDES permits for storm sewers (and irrigation point source discharges) was administratively infeasible. The EPA contended that the permit program required establishing national numeric effluent standards. Since the owner of a discharge point has little control over the quantity of flow or the nature and amounts of pollutants picked up by the runoff, promulgating meaningful effluent limitations would be impossible. The court, however, held that *numeric* effluent limitations are not the only permissible limitations on a discharge, quoting section 302(a) of the Clean Water Act which expands effluent limitations to include *alternative effluent control strategies*.

Therefore, the EPA has the flexibility, when numerical limitations are infeasible, "to issue permits with conditions designed to reduce the level of effluent discharges to acceptable levels. This may well mean opting for a gross reduction in pollutant discharge rather than the fine-tuning suggested by numerical limitations" (*Natural Resources Defense Counsel, Inv. v. Costle*). It should be pointed out here that this decision also applied to point sources discharge from irrigated agriculture. The irrigation discharges were thereafter specifically exempted from the permit program by the 1977 amendments to the act.

What these conditions, or alternative effluent control stategies, should be is not addressed in the new EPA regulations. Obviously, a storm sewer is not a waste treatment plant which can add more treatment capabilities to produce a better effluent. Rather than applying end-of-pipe pollution control technology, the conditions in a storm sewer permit should control the pollution at it sources. The attempt must be made to control various pollutants from entering the storm sewer in the first place. The authority to impose such controls, which might include erosion control, street sweeping, and prevention of oil and grease discharge from service stations and car washes, may lie with different levels of government; or the authority may still need implementing regulations or ordinances. Section 208 agencies may develop "best management practices" which can be incorporated into general permits.   The point is that the act, as interpreted by the courts, gives water quality administrators adequate flexibility in the permit program to utilize many different appropriate techniques.

Control of Stormwater Runoff (Nonpoint Source Pollution)

Section 208 of the act requires the development of areawide wastewater man-agement plans throughout the nation.   They are being prepared by designa-ted local agencies (such as regional councils of government) and by state agencies.   The purpose of section 208 is to develop 20-year plans for clean water which would form the basis for many of the other programs of the act. Stormwater runoff is specifically mentioned as a pollution source which must be addressed in a section 208 plan, together with other nonpoint sources of pollution.   The plans are also supposed to identify recreation and open space opportunities.

Nonpoint sources of pollution are to be controlled by "best management practices" developed by section 208 agencies in cooperation with other appro-priate agencies and persons.   These practices can take many forms, and may require the cooperation of different levels of government and private entities. Such cooperation can result in improving not only water quality, but other public benefits as well.   For example, keeping a floodplain or natural drain-ageway undeveloped may not only be wise floodplain management providing for open space and recreation, but could also aid in cleansing urban storm-water run-off as it flows through grasses and other vegetation on its way to the stream.   In addition, drainage facilities should be planned and designed with water quality objectives in mind.   Pollution from urban runoff can be aggravated by installing storm sewers which quickly flush waste and debris from streets and other impermeable surfaces into streams.   Using a different strategy for urban drainage, on the other hand, could satisfy both drainage and water quality objectives.   A section 208 plan is a useful vehicle for inte-grating these objectives.

State and Local Laws

Most states have enacted their own water quality legislation and have also taken over the NPDES permit program.   States establish their own water quality standards and classify their streams accordingly.   These may affect the effluent limitations placed in the NPDES discharge permits.   States are also actively involved in the section 208 plans and their certification.   In addition, a state or local government may have adopted special regulations re-garding erosion control or other aspects of storm runoff quality.   Therefore, drainage planners must also conform to state and local water quality regula-tions.

Conclusions

It is essential that drainage planners work closely with the water quality agencies throughout the planning process.   This will not only assure the necessary compliance with all applicable federal, state, and local water quality laws, but result in the implementation of the goal of clean water.   While single-purpose drainage facilities may result in additional water pollution, in-tegration with water quality planning may result in multipurpose facilities

which are more cost-effective and could attain other community goals for
recreation and open space.

2.6 FINANCING THE PROJECT: THE DRAINAGE AND FLOOD CONTROL UTILITY FEE

Communities have long found it difficult to finance drainage projects. It is
not the intent here to discuss an array of funding possibilities, but to focus
on one innovative solution which has been successfully implemented. In an
effort to equitably finance its drainage plan, the city of Billings, Montana
decided to view drainage facilities as part of a drainage utility and charge for
the services provided. Property owners whose runoff drained into city storm
sewers and drainage facilities would be considered customers of the storm
sewer utility, just like citizens whose homes used city water and sewer facilities. The fee charged would essentially be based on the difference between
historic runoff and the amount of runoff from that property in its developed
state. The reasoning was that under natural conditions a considerable amount
of stormwater percolates into the ground. However, where land is covered
with homes, carports, parking lots, and other impermeable surfaces, much
more water runs off, at greater velocities, causing higher peak flows than
naturally. Flat rates would be charged for different types of residential property; and commercial establishments, which usually have more impervious surfaces than residential properties, would be charged a higher rate. The proposal was challenged in *City of Billings v. Nore*, 148 Mont. 96, 417 P.2d 458
(1966). The proposal was upheld as constitutional and equitable and has
since been implemented. Boulder, Colorado, has also adopted and implemented
the drainage utility and fee concept.
    Additional refinements can include:

> Giving credit for on-site detention. Since the amount of runoff will be
>   less, the drainage fee is reduced. Giving credit is an incentive to
>   on-site storage, which keeps runoff as close to natural as possible,
>   thereby safeguarding against potential liability for increasing or con-
>   centrating surface waters. However, since there will always be a
>   need for providing and maintaining the basic citywide drainage and
>   flood control facilities which benefit all citizens, all properties should
>   be charged a basic fee, even where the historic flows from that par-
>   ticular site are being maintained.
> Providing that drainage fee revenues may be used not only for structur-
>   al projects, but also for nonstructural measures such as purchase of
>   land or easements to preserve natural drainageways.
> Providing for the option of calculating the amount of actual runoff from
>   a particular parcel, such as a shopping center, in order to more
>   precisely determine the fee.
> Adding a surcharge to the drainage fee for developed properties situ-
>   ated in a floodplain or flood hazard area because of the extraordinary
>   public costs involved in protecting the properties and in providing
>   emergency services in the event of a flood.

    A drainage plan is of little value unless it is implemented. Corrective
measures are costly. Financial resources are needed for both corrective
measures and preventive measures such as purchase of natural drainageways
in advance of development. A drainage fee is an equitable and legal
method of providing such financial resources.

2.7 SUMMARY AND CONCLUSIONS

1. Drainage planners should be aware of the principles of stormwater law in order to avoid legal obstacles to implementation of plans, to minimize potential liability, and to be alerted to opportunities.

2. Legal principle may determine which alternative among a number of viable engineering alternatives should be chosen.

3. The law distinguishes between watercourses and surface waters. A watercourse has definite bed and banks, while surface waters run in a diffused manner overland.

4. The law of watercourses is based on the riparian doctrine which holds that each riparian owner has an equal right on the stream; he may protect himself from ordinary floodwaters, but only if his actions do not injure other riparian owners.

5. Three different doctrines have evolved in surface water law:

Common enemy rule. Under the pure common enemy rule, both the upper and lower property owners could protect themselves from surface waters even if it harmed the other's property. Where jurisdictions have adopted this rule, one must analyze how they have modified it. Some use an overriding principle that one's actions in protecting one's properties from the enemy cannot make that enemy more dangerous to one's neighbor.

Civil law rule. Under the civil law rule, the upper landowner has a drainage easement over the lower property, and the lower property must take such drainage. However, the easement is for natural drainage. While the drainage may be altered, liability may result if surface waters are sent down in a manner, quantity, or location to do more harm than formerly.

Reasonable use. The reasonable use rule recognizes the right of each owner to deal with his property as he wishes, but he must act reasonably under all circumstances.

6. The distinctions between watercourse law and surface water law, and among the three rules of surface water law, are important. However, since courts may establish new principles and duties and apply them retroactively to the case before them, legal counsel should give advice which would protect the drainage planning entity from liability under any of these principles.

7. Natural drainage conditions may be altered, but liability may result if such alterations cause damage. Alterations include reducing the permeability of the land, embanking, damming, diking, channelizing, changing the location, increasing the volume or the velocity, increasing the drainage area, concentrating, raising the channel, and adding debris.

8. Not even the extraordinary storm, or act of God, has been held to be an absolute defense. Parties have been held liable for damages where their action, concurring with the act of God, was the proximate cause of the damage. In addition, with improved technology and better data, storms and floods are now categorized according to their statistical probability. The concept of the unanticipated or extraordinary storm will become less and less available as a defense against liability.

9. Some courts hold that the flood of record, regardless of size, is not an extraordinary flood, and that when a new flood of record occurs, this becomes the new standard to be met.

10. Where facilities have been built, however, which accommodated more than double the flood of record at the advice of nationally known authorities, and the facilities were unable to pass an even greater flood, the city was held not liable.

11. Drainage problems should be solved within a basinwide planning strategy. Ad hoc or partial corrective solutions can cause more problems than they solve.

12. The best approach in planning and designing drainage works is to attempt to retain natural and historic conditions of flow.

13. Drainage planning should be based on runoff which will result from future urban development which can be reasonably anticipated.

14. Cities and counties should require that new urban development not materially increase the amount of storm runoff so as to alter drainage conditions. This is a legitimate use of the police power to safeguard those downstream and to protect the urban infrastructure. It would also minimize the potential liability to the city or county which may result from issuing permits for buildings or approving a subdivision plot which causes harm.

15. On-site detention of stormwaters should be encouraged, not only because it decreases the size and therefore the cost of storm sewers, but also because it is a safeguard against potential liability for concentrating or increasing surface water runoff.

16. Whenever possible, artificial channels should follow natural thalwegs.

17. Transbasin diversions which increase natural flow should be avoided unless the risks are adequately evaluated and protected against.

18. Installation of inadequately sized drainage and flood control structures should be avoided, especially if such structures cause development and filling of the natural watercourse so that larger flood flows are altered, causing damage to properties which would not have been damaged otherwise.

19. The 100-year (1 percent) flood may be the appropriate regulatory or design criterion for some purposes. A larger flood of record may be the determining criterion in some states. On the other hand, drainage facilities which are built for convenience purposes rather than protection of life and property may be designed for less stringent criteria.

20. Because of the increased flooding potential, an embankment constructed to detain or retain water should be safe from failure in the event of larger floods. The maximum probable flood would be a prudent criterion.

21. Subdividers and developers are treated like individual upper and lower property owners in surface water cases and like individual riparian landowners in watercourse cases. Therefore, they are subject to the same potential liability. This liability is based on the common law and is independent of any subdivision regulations with which they may or may not have complied.

22. Under the common law a public entity has no duty to provide drainage and flood control facilities. Having done so, however, public entities are treated like private parties when their facilities cause damage.

23. Public entities have been found to be liable for their negligence, have been held strictly liable (that is, without negligence), and have been held liable for violation of surface water rules and watercourse law.

24. When public entities have built facilities which caused intermittent or permanent flooding of private land, they have had to reimburse landowners for "constructively taking" such land.

25. Some courts are also finding cities liable for issuing permits for private developments which cause injury to other properties.

26. Governmental immunity has rarely been a successful defense against liability in drainage cases.

27. Potential liability under stormwater law for public and private entities, their engineers, and planners must be considered in choosing between viable engineering alternatives.

28. Floodplain and wetland regulations are legitimate tools to protect public health, safety, and welfare. Courts are recognizing that such regulations:

> Prevent development in flood hazard areas which could be destroyed or damaged and cause loss of life and property
>
> Protect the unwary buyer from victimization
>
> Protect properties upstream, downstream, and laterally from obstructions in the floodplain which would raise flood heights, divert waters, or increase velocities
>
> Preserve the flood-carrying capacities and transient flood storage capacity of the floodplain
>
> Prevent utilities and other components of the urban infrastructure from being built in hazardous locations
>
> Preserve the groundwater recharge capabilities of the floodplain
>
> Avoid the need for public expenditure for flood protection works such as dikes and channels and public monies for disaster relief
>
> Prevent community disruption and economic losses
>
> Protect the tax base
>
> Protect water quality, aquatic life, wildlife, and natural ecosystems

29. Nonstructural floodplain management results in a program which has the least exposure for the community in terms of potential liability.

30. The National Flood Insurance Program has minimum requirements which must be met for a community and its citizens to be able to participate in the program. These include adopting regulations for management of the 100-year (1 percent) flood. States or local communities may adopt more stringent requirements than the federal minimum.

31. Drainage planners should work closely with water quality planners. Pollution control through management of urban stormwater runoff should be an integral part of drainage planning in order to attain the goal of clean water and other community goals regarding recreation and open space. The Area-wide Wastewater Management Plans (section 208 plans) now being undertaken nationwide are good vehicles for such integration.

32. Urban stormwater runoff can be either a point source or a nonpoint source of pollution under the federal Clean Water Act, depending on whether it enters a stream from a storm sewer or from diffused overland flow.

33. Discharges into and from storm sewers can be controlled by National Pollution Discharge Elimination System (NPDES) permits, while nonpoint sources are to be controlled by "best management practices." It is essential that drainage planners take such present or potential permits and practices into account.

34. Most states have adopted their own water quality legislation and taken over the NPDES permit program. Local regulations affecting water quality and stormwater runoff may also have been promulgated. It is essential that drainage plans also conform to these state and local laws.

35. A drainage and flood control utility fee has been found to be an equitable and legal method of financing drainage facilities.

CASES

Armstrong v. Francis Corporation, 20 N.J. 320, 120 A.2d 4 (1956).
Ayala v. Philadelphia Board of Public Education, 453 Pa. 584, 305 A.2d 877 (1973).
Barr v. Game, Fish and Parks Commission, 30 Colo.App. 482, 497 P.2d 340 (1972).
Bartlett v. Zoning Com'n of Town of Old Lyme, 161 Conn. 24, 282 A.2d 907 (1971).
Beckley v. Reclamation Board of State, 205 Calif.App.2d 734, 23 Calif.Rptr. 428 (1962).
Beetison v. Ballou, 153 Nebr. 360, 44 N.W.2d 721 (1950).
Biffer v. City of Chicago, 278 Ill. 562, 116 N.E. 182 (1917).
Board of Commissioners of Logan County v. Adler, 69 Colo. 290, 194 P. 621 (1920).
Brecciaroli v. Connecticut Com'r of Env. Protec., 168 Conn. 349, 362 A.2d 948 (1975).
Breiner v. C & P Home Builders, Inc., 398 F.Supp. 250 (E.D.Pa. 1975), aff'd in part, rev'd in part 536 F.2d 27 (3rd Cir. 1976).
Cabral v. City and County of Honolulu, 32 Hawaii 872 (1933).
Candlestick Prop., Inc. v. San Francisco Bay C. & D. Com'n, 11 Calif.App.3d 557, 89 Calif.Rptr. 897 (1970).
Cappture Realty Corp. v. Board of Adjustment, 126 N.J. Sup. 200, 313 A.2d 624 (1973).
Chicago & Alton R.R. v. Tranbarger 238 U.S. 67 (1915).
Chicago, R. I. & P. Ry. Co. v. Groves, 20 Okla. 101, 93 P.755 (1908).
Chicago, R. I. & P. Ry. Co. v. Taylor, 173 Okla. 454, 49 P.2d 721 (1935).
Chudinsky v. City of Sylvania, 372 N.E.2d 611 53 Ohio App.2d 151 (1976).
City of Billings v. Nore, 148 Mont. 96, 417 P.2d 458 (1966).
City of Chicago v. Washingtonian Home of Chicago, 289 Ill. 206, 124 N.E. 416 (1919).
City of Houston v. Wall, 207 S.W.2d 664 (Tex. Civ.App. 1947).
City of Pascagoula v. Rayburn, 320 So.2d 378 (Miss. 1975).
City of Vicksburg v. Porterfield, 164 Miss. 581, 145 So. 355 (1933).
City of Welch v. Mitchell, 95 W.Va. 377, 121 S.E. 165 (1924).
Davis v. Fry, 14 Okla. 340, 78 P. 180 (1904).
Dayley v. City of Burley, 524 P.2d 1073 96 Idaho 101 (1974).
Elliott v. Nordlof, 83 Ill.App.2d 279, 227 N.E.2d 547 (1967).
Frederick v. Hale, 42 Mont. 153, 112 P. 70 (1910).
Godlin v. Hockett, 272 P.2d 389 (Okla. 1954).
Goldblatt v. Hempstead, 369 U.S. 590 (1962).
Hankins v. Borland, 163 Colo. 575, 431 P.2d 1007 (1967).
Iowa Natural Resources Council v. Van Zee, 261 Iowa 1287, 158 N.W.2d 111 (1968).
Just v. Marinette County, 56 Wis.2d 7, 201 N.W.2d 761 (1972).
Livingston v. McDonald, 21 Iowa 160, 89 Am.Dec. 564 (2d syllabus) (1866).
MacGibbon v. Board of Appeals of Duxbury, 347 Mass. 690, 200 N.E.2d 254 (1964), 356 Mass. 696, 255 N.E.2d 347 (Mass. 1970).
McManus v. Otis, 61 Calif.App.2d 432, 143 P.2d 380 (1944).

Masley v. *City of Lorain*, 48 Ohio St.2d 334, 358 N.E.2d 596, (1976).
Miller v. *Bay Cities Water Co.*, 157 Calif. 256, 107 P. 115 (1910).
Miller v. *City of Brentwood*, 548 S.W.2d 878 (Tenn.App. 1977).
Missouri, K. & T. Ry. Co. v. *Johnson*, 34 Okla. 582, 126 P. 567 (1912).
Morris County Land I. Co. v. *Parsippany-Troy Hill Tp.*, 40 N.J. 539, 193 A.2d 232 (1963).
Mozzetti v. *City of Brisbane*, 67 Calif.App.3d 565, 136 Calif.Rptr. 751 (1977).
Myotte v. *Village of Mayfield*, 54 Ohio App.2d 97, 375 N.E.2d 816 (1977).
Natural Resources Defense Counsel, Inc. v. *Costle*, 568 F.2d 1369 (1977).
Natural Resources Defense Counsel, Inc. v. *Train*, 396 F.Supp. 1393 (D.C. Cir. 1975).
Oklahoma City v. *Bethel*, 175 Okla. 193, 51 P.2d 313 (1935).
Oklahoma City v. *Evans*, 173 Okla. 586, 50 P.2d 234 (1935).
Oklahoma City v. *Rose*, 176 Okla. 607, 56 P.2d 775 (1936).
Oklahoma Ry. Co. v. *Boyd*, 140 Okla. 45, 282 P. 157 (1929).
Pennsylvania Coal Co. v. *Mahon*, 260 U.S. 393 (1922).
Potomac Sand & Gravel Co. v. *Governor of Maryland*, 266 Md. 358, 293 A.2d 241 (1972), cert. denied 409 U.S. 1040.
Pumpelly v. *Green Bay Co.*, 80 U.S. 166 (1871).
Rivers Defense Committee v. *Thierman*, 380 F.Supp. 91 (S.D. N.Y. 1974).
Sands Point Harbor, Inc. v. *Sullivan*, 136 N.J.Super. 436, 346 A.2d 612 (1975).
Sibson v. *State*, 115 N.H. 124, 336 A.2d 239 (1975).
Spiegle v. *Borough of Beach Haven*, 116 N.J.Super.Ct. 148, 281 A.2d 377 (1971).
State v. *Deetz*, 66 Wis.2d 1, 224 N.W.2d 407 (1974).
State v. *Johnson*, 265 A.2d 711 (Me. 1970).
State v. *Reed*, 78 Misc.2d 1004, 359 N.Y. S.2d 185 (1974).
State Highway Commission v. *Biastoch Meats, Inc.*, 145 Mont. 261, 400 P.2d 274 (1965).
Town of Jefferson v. *Hicks*, 23 Okla. 684, 102 P.79 (1909).
Turner v. *County of Del Norte*, 24 Calif.App.3d 311, 101 Calif.Rptr. 93 (1972).
Turnpike Realty Company v. *Town of Dedham*, 362 Mass. 221, 284 N.E.2d 891 (1972), cert. denied 409 U.S. 1108 (1973).
United States v. *Lewis*, 355 F.Supp. 1132 (S.D.Ga. 1973).
United States v. *Joseph G. Moretti, Inc.*, 331 F.Supp. 151 (S.D.Fla. 1971).
Vartelas v. *Water Resources Commission*, 146 Conn. 650, 153 A.2d 822 (1959).
Vazza Properties, Inc. v. *City Council of Woburn*, 296 N.E.2d 220 (Mass.App. 1973).
Village of Euclid v. *Ambler Realty Co.*, 272 U.S. 365 (1926).
Zabel v. *Tabb*, 430 F.2d 199 (5th Cir. 1970), certiorari denied 401 U.S. 910 (1971).

3
The Drainage Planning Process

3.1 INTRODUCTION

Comprehensive drainage planning is required to achieve the goals and objectives of an urban region. The communitywide planning process views drainage as a subsystem of a larger and more important urban system. Thus drainage and flood control efforts are a means to an end and not entirely ends in themselves. A given drainage plan may be oriented toward various portions of the drainage network, but should always address important overall development and regional resource management relationships.

The scope of each planning step described herein should be relative to the significance of the factors involved, such as existing problems, land uses, and opportunities to integrate drainage into more comprehensive program efforts. As a study progresses, the need to further investigate data and refine planning work will usually occur. Thus, the planning process is an iterative process.

Products

The drainage planning process can culminate in a variety of reports. These include floodplain information reports, conceptual reports, planning reports, and preliminary design reports. The information presented will relate to:

> Identification of the basin characteristics
> Hydrologic simulation of a drainage basin which will reasonably represent existing and future land use conditions and identify the probable results of management strategies
> Delineation of floodplains and flood problems
> Identification of the range of management alternatives, including structural and nonstructural components
> Description of the effects of alternatives in terms of costs and goal achievements

Overall Results of the Drainage Planning Process

The planning process should produce the following results:

> Drainage planning becomes an input into overall development strategies for the region. The runoff network is recognized as a subsystem of a broader urban system.

35

The drainage program becomes an active, positive program which assists in the implementation of overall goals rather than a reactive program which addresses drainage and flood control problems after they become apparent.

Drainage management strategy results in multiple use of the floodplain resource and long-term cost-saving programs.

3.2 PROCEDURES

There are four major actions which have to be taken in order to complete a drainage plan:

Field surveys and library research
Analysis
Reporting
Implementation

The depth to which each action is investigated depends upon the resources available and the importance of the action to the overall process. Therefore, the planning process is generally iterative and all of the actions are covered a number of times so that the more critical ones are investigated to greater depth.

The final measure of success of the plan is dependent upon the degree to which it is implemented. Thus, it is important that three actions occur as part of implementation:

Flood-prone areas should be designated and regulated.
There should be an ongoing process of interaction between developers, their consultants, and the representatives of various governmental bodies.
The plan should be followed by all parties concerned.

3.3 FIELD SURVEYS AND LIBRARY RESEARCH

The field survey and library research stage in the planning process is necessary to acquire data to be used in rainfall-runoff hydrology and conceptualization of alternatives. Descriptions of the required data follow.

Location Maps

Location maps should describe the general physical location of the basin of concern in relation to other major watercourses, various political boundaries, and major physical features.

Topography

Maps depicting the relief stream network development are necessary. Detailed topographic maps showing elevation contours in good detail are desirable.

Vegetation

Aerial photographs or various inventory maps that describe the kinds and quantities of vegetation in the basin are needed to reliably depict the runoff process.

Soil Survey

A soil survey is important to understand the infiltration and runoff process in a basin. Information is needed relating to the physical condition of the soil such as structural stability and erosive tendencies.

Geology

Data or maps are needed to describe surface and bedrock geology. Hydrogeologic information is also needed to reliably describe the runoff response of a basin.

Stream Survey

Cross sections, either from the topographic maps or from field instrument surveys, are needed for hydrologic stream routing and hydraulic calculations. For the same reasons, data relating to channel conditions such as roughness, vegetation, and meandering are needed.

Existing Development

Existing development data are needed to describe land use, transportation, water features, and waterway characteristics. This information will be used in other areas of the planning process such as runoff hydrology and alternative conceptualization.

Future Development

Land use forecasts must be acquired to the greatest detail available. Usually maps that depict areas of existing and proposed development are available from the planning authorities.

It is also important to review development regulations, as the structures and landscape constructed to comply with these regulations can significantly control the runoff response of a basin. One of the results of a drainage planning study can be to significantly modify land use or development regulations.

Rainfall

Drainage planning involves the recognition of the probabilities of certain events occurring in order to identify the best management methods. Drainage systems are usually designed based on a precipitation event which has a certain statistical frequency of occurrence. However, proposed systems are usually evaluated for performance during a range of other events depending upon the importance of a system.

Precipitation data are usually available in raw and synthesized formats according to which rainfall-gaging station is being used. Information regarding frequency, duration, spatial relationships, and temporal relationships is generally available. This information should be used with good judgment so that unusual critical relationships do not escape the evaluation. Chapter 4 on rainfall provides much of the pertinent information for use in most planning efforts.

An assumption is made throughout this book that a given frequency rainfall of a critical duration will lead to the same frequency runoff flood (i.e., the 100-year rainfall results in the 100-year flood). It shoud be recognized that this is not the exact case. Different rainfall events can lead to runoff events which do not have the same frequency of occurrence. Present hydrologic tools and data available are still too limited to make practical comprehensive analysis of this situation.

Drainage Management Strategy

Drainage management strategy as envisioned by the concerned parties must be inventoried so that possible constraints and positive input are recognized early in the planning study.

3.4 ANALYSIS

This portion of the planning process will require an iterative approach in which each step reveals new information and thus points out the possible need for further analysis of previous work. The major steps included herein are:

    Runoff analysis
    Floodplain delineation
    Problem inventory
    Recognition of alternative components
    Recognition of multipurpose opportunities and constraints
    Conceptualization of alternatives
    Hydrologic analysis of possible alternatives
    Cost/benefit analysis and evaluation

Runoff Analysis

Runoff analysis establishes the important baseline conditions by describing in numbers the runoff character of the drainage basin. Chapter 5 on runoff describes several methods available.

Rainfall Excess
The information in Chap. 4 on rainfall is a primary input into the hydrologic analysis. A certain amount of this rainfall will always be lost to abstractions, primarily infiltration and surface detention losses. It is important to describe the rainfall excess for existing basin conditions and future possible basin conditions since dramatically different runoff can occur.

Subbasin Delineation
A major step in runoff hydrology is to identify subbasins which take into consideration the following major points:

    Major tributaries that have significantly different characteristics, or for
        which discharge flow information is desired
    Areas of existing development
    Areas of proposed development
    Possible floodwater storage sites
    Reaches of stream to which significant alterations may be made in the
        future

Runoff Conceptualization

An important role of the hydrologist is to identify the key runoff phenomena that are occurring and to represent these with reasonable mathematical simplification. Accordingly, he must select an appropriate runoff model, input appropriate data, and select a routing scheme.

Floodplain Delineation

The flood hydrographs that are produced in the runoff step are then used as input to a hydraulic analysis procedure that identifies the floodplain for a given flow. The array of techniques that can be used to predict water surface profiles is explained in Chap. 7 on floodplain delineation. As well as the runoff flows, there are three basic types of data necessary for floodplain delineation. These are:

Representative cross sections of the stream or channel
Information regarding the hydraulic character of the sections and their
relationships, such as roughness, stream slope, and meandering
Hydraulic information regarding bridges, culverts, and other constrictions or controls which can create a different water surface than that
which would be caused by the cross section alone

The analysis of this information will result in a profile (or longitudinal section of the stream) and a plain view-plot which together depict the extent of the floodplains for a given flow. This information, when combined with other analysis information regarding depths of flow and velocity, can characterize the extent of probable damages.

Problem Inventory

One of the most important steps in the planning process is to identify and describe quantitatively and qualitatively the existing and potential drainage problems in a basin. There are several actions that are usually necessary as part of the problem inventory:

Concerned municipalities, individuals, and special technical groups
should be interviewed to understand their view regarding drainage
problems. It is necessary to identify the symptoms that are
discussed in relation to the actual problems.
The floodplains delineated for existing conditions and possible future
alternatives should be carefully analyzed to assess the scope of
potential damages, to identify existing and future problems, and to
also identify certain advantageous situations which can be a key to
preventing or relieving problems.

Alternative Components

There are many possible alternative components that could be used to correct or prevent drainage problems. These components are usually used in combinations but in some cases are used on an individual basis. The components can be categorized in many ways, but they are identified herein as non-structural and structural components. Nonstructural components seek to *mitigate the effects of a flood* and are an approach that recognizes the floodplain as nature's prescribed easement. Structural components seek to *affect the flood event* by changing its distribution in time and space.

Nonstructural Components
Nonstructural components generally keep development from occurring in the floodplains. Further, they inform existing development of flood hazard and suggest mitigating actions that would limit the damages incurred during a flood event.

Nonstructural components include:

Delineation of floodplains
Control of floodplain land uses
Acquisition of selected floodplains
Subdivision regulations
Floodplain information and education
Flood forecasts and emergency measures
Flood proofing
Flood insurance (if available)

These components are further elaborated in Chap. 8 on nonstructural planning.

Structural Components
Various constructed works can be used to store or convey floodwaters to reduce damages. There are many potential structural alternatives and many important details required for each. A complete explanation of these components and their requirements is beyond the scope of this text. Readers are referred to the textbooks and source documents regarding hydraulic structures which describe potential structural components, design, analysis, and other important considerations.

Listed below are four general types of structural components:

Channels. These include numerous different possible solutions such as hard concrete-lined channels, soft grass-lined channels, use of natural channels with some clearing and erosion control works, and hybrid solutions such as soft-lined bottoms with retaining walls.
Pipes and conduits. These are usually used in the upper reaches of streams and may be precast or cast-in-place structures.
Constriction removal. This can be exemplified by the replacement of various bridges and culverts to allow design flows to pass without undue backwater effects.
Man-made storage. This storage is used to supplement the natural storage. There are three possible types of storage: detention storage, retention storage, and conveyance storage. These are dealt with in detail in Chap. 10 on man-made storage planning.

Multipurpose Planning Opportunities and Constraints

Drainage planning should incorporate compatible multipurpose planning concepts and recognize constraints created by other community facilities. The following paragraphs cite examples.

Recreation
Recreation opportunities can relate to the waterway and/or to the floodplain.
Water-oriented recreation includes boating and fishing. Recreation uses of
the floodplains include sports fields, picnic areas, and space and trails for
walking, hiking, biking, and horseback riding.

Such recreation opportunities can result from a drainage program.
However, it must be realized that these features may not be usable during a
flood event and may suffer some flood damages.

Solid Waste Disposal
As a catchment develops, there is increasing demand to fill in the floodplain,
both to acquire land and because there is need for space for solid waste.
When solid waste disposal sites are not provided at convenient locations, what
often happens is that this material ends up in the streams and waterways of
the basin. Solid waste disposal sites can be planned in conjunction with
drainage planning.

Transportation
The majority of drainage basins and their stream networks are significantly
changed by transportation facilities, such as freeways, local roads, railways,
and airports. Combined efforts of transportation and drainage planning can
often result in improved facilities at a lesser cost. When planned, necessary
fill for road embankments can come from drainage work construction; yet
when unplanned, one can find instances where the carrying capacity of natural waterways has been severely limited by filling in the floodplain. Often
new roads for a subdivision can be planned to provide embankments for detention and retention storage sites.

When drainage and transportation planning are done together, the inconvenience to traffic movement can be minimized during severe flooding events.
On a larger scale, transportation facilities, such as freeways, can be planned
to provide major flood storage facilities.

Land Use
Many communities desire open space and recreational opportunities which enhance the quality of life. Drainage planning can assist in the attainment of
such opportunities.

Land development is of paramount importance to the drainage system
because of the increased runoff due to urbanization. Appropriate shifts in
land use patterns can reduce potential drainage problems and alleviate expenditures on drainage works.

Water Supply
Water for irrigation, industry, and other purposes can be obtained in conjunction with drainage planning, particularly with regard to man-made
storages. Also, opportunities exist for improving groundwater availability.
For example, a strategically located retention basin can provide water to recharge an aquifer that is being overpumped.

Water Quality
Runoff water quality during rainfall events can be poor. There are many
possibilities to reduce water quality degradation in a drainage program. Such

possibilities include designation of on-site drainage and land use practices which minimize the source of pollutants, sedimentation and debris basins, aerated ponds, and land treatment (including soil infiltration and grass filtration) methods.

Extractive Industry
Extractive activities which remove sand, clay, and rock for building purposes can be planned and controlled so that the exhausted sites can be developed as storage facilities for floodwaters.

Conceptualization of Alternatives

Conceptualization of alternative drainage plans is another major step in drainage planning. It is a challenging process because of the nearly unlimited number of component combinations. There is a strong need to come up with a feasible number of alternatives for analysis and the final designation of one alternative. Thus, judgment is always required in the process of delineating the alternatives. The following paragraphs describe a simplified procedural outline for conceptualization of alternatives.

Common Items
Part of any drainage plan are common items and points which need to be addressed:

> *The floodplain areas should be delineated and information disseminated on the hazard.* Chapter 9 on floodplain regulations explains this action in more detail. The flood-prone area should be based upon the expected future development and present stream conditions. In this way, the introduction of development resulting in creation of significant potential damages can be prevented. If subsequently the basin develops differently, and/or different drainage works are undertaken, the flood-prone area should be modified to suit the new conditions.
> *Land use planning input.* Because land use has a significant effect on the drainage response of a basin, it is important to incorporate existing and proposed land uses. Land use planners must be made aware of the drainage impacts of development and must seek to modify development patterns so as to prevent the creation of drainage problems.
> *Interaction.* Drainage planning is a dynamic process. Previously unrecognized problems, flexibilities, and constraints (as anticipated from other agencies, local communities, and persons directly concerned) must be integrated into the planning process. It is most important that this information be made available to appropriate interest groups during the planning process.

Evaluation of Possibilities
The list of alternative components previously presented should be thoroughly evaluated. The following general procedure should be used:

> Recognize and identify combinations of components.
> Relate to overall goals and objectives.
> Assess and eliminate unreasonable combinations.
> Identify the most likely combinations; the least likely combinations
>   should be held aside in case it becomes apparent that the more likely

combinations have serious problems, or if there should be other reasons for including such alternatives at a later time.
Identify the major management areas within the basin which are of critical importance to the drainage system.

With this process, the basic alternatives are identified along with the major areas of activities. Again, this planning process is an iterative process in which the earlier steps may be readdressed in more detail as areas of concern are exposed in the later steps. For example, in the conceptualization step it may become apparent that it is necessary to acquire more data and identify possible problem areas more thoroughly.

Hydrological Analysis of Possible Alternatives

The procedures outlined under "Runoff Analysis," above, should be followed again for each of the possible alternatives. This would include runoff routing and preparation of runoff hydrographs which will show volumes and peak discharge values in comparison with the existing conditions and future conditions with no modifications to the stream network.
It is important at this point to analyze these results to identify both the negative and positive modifications made to the runoff response characeristics of the basin with various alternatives.

Cost/Benefit Analysis

The analysis of the costs and benefits of each of the alternatives is a demanding process requiring consistency, fairness, and reliability.
All evaluations should be made for the total system required for a given reach in the drainage network. The reaches should be chosen so that the evaluations are equal and fair to all alternatives.
The depth of this analysis should be scaled appropriately to the magnitudes and complexities of the basin and study level involved. This depth has to be chosen with careful judgment and should be agreed upon by the principal study participants.

Alternative Cost Estimate
There are two major areas that need to be addressed to assess the cost of alternatives. They are capital costs for construction and the operation and maintenance costs. Some of the components of each of these are presented below:

Capital
Pipes
Channels
Erosion protection
Embankments
Structures
Clearing and removal of obstruction
Land acquisition

Operation and maintenance costs
Pipes
Channel repairs
Erosion protection
Mowing grass and other vegetation maintenance

Structure (including embankments) repairs
Clearing of troublesome sediment, debris, and obstructions
Inspection of waterways or facility conditions
Inspection of detention, retention, and flood-prone areas

Inventory of Benefits
This is a difficult area of analysis because many of the benefits, particularly those related to environmental and social aspects, cannot be expressed readily in dollars.   Attempts at expressing the benefits in dollars can be a guide, but certainly are easily questioned. The depth to which the benefits ought to be inventoried should be scaled to the magnitude, complexity, and expense of possible alternatives. There are several factors which would normally be covered in any planning report. These are:

Damages and damages relieved. Damage assessments should usually be cursory for most cases, since measurement of damages due to an expected water surface is speculative. The U.S. Army Corps of Engineers, U.S. Flood Insurance Administration, and U.S. Soil Conservation Service, along with other groups, have procedures that can provide potential damage information for relative water levels of certain structures [1-3].

Other economic factors. These would include approximate assessments for inconvenience due to such things as transportation disruption.

Multiple-use benefits. Assessing these includes an evaluation of the multipurpose planning opportunities. When possible, economic benefits and debits should be assigned.

Environmental factors. These include assessments of the impacts of the natural ecosystem.

Social factors. These include consideration of peace of mind to community citizens when potential drainage problems are reduced. Assessments of impacts on the quality of life should also be made.

Drainage Law Considerations

As part of the planning process, it is important to identify the legal and administrative constraints which impact on drainage. It is usually wise to include a legal professional as part of the planning process.

3.5 REPORTING

A key element to the usefulness and success of a plan is an understandable and comprehensive report. This report should be used as part of the interaction process to refine and select the alternative plan. There are two important components of this process which are described in this section.

Display of Alternatives

The report should include appropriate written material, graphs, tables, figures, and maps which present the major study points in summary form. These could include:

Overall and subbasin maps
Figures and maps which depict in detail the various alternatives for important areas within the basin

Economic, environmental, and social impacts displayed in an
understandable manner with sufficient details to facilitate decisions
as to the preferred alternatives

Evaluation and Refinement

As part of this reporting process, the alternatives should be presented in a
manner that will promote clear understanding of the range of impacts associated with the alternative proposals. This will allow interaction with local
authorities and interests when selecting the preferred alternative.

REFERENCES

1. *Guidelines for Flood Insurance Studies*, U.S. Army Corps of Engineers,
   Washington, D.C., January 1970.
2. *Flood Damage Estimates—Residential and Commercial Property*, U.S.
   Department of Agriculture, Soil Conservation Service, WEP Technical
   Guide No. 21, Supplement 1, South Regional Technical Service Center,
   Fort Worth, Tex., December 1970.
3. *Flood Hazard Factors, Depth-Damage Curves, Elevation Frequency
   Curves, Standard Rate Tables*, U.S. Federal Insurance Administration,
   Washington, D.C., September 1970.

4
Rainfall

4.1 INTRODUCTION

This chapter on rainfall presents information on the physical phenomena of rainfall, basic approaches that are used to analyze and generate rainfall statistics throughout the world, and examples of how these statistics are used for specific locations.

Rainfall statistics are a fundamental input to runoff hydrology. Typically, events are analyzed in terms of return period and event duration.

The return period, or recurrence interval, is defined as the average interval in years between the occurrence of events of a certain size. Its probability of occurrence is defined as the reciprocal of the return period.

Rainfall data are also analyzed in terms of duration of a particular event. Generally smaller basins are more sensitive to shorter durations while the larger basins are more sensitive to the longer durations.

Usually several return period and duration events are analyzed in drainage planning to develop an understanding of the range of runoff results. Specific events, such as the 10- and 100-year return period rainfall runoff, are used for design purposes.

Rainfall Phenomena

A brief explanation of rainfall phenomena is appropriate in developing an understanding of the cause of rainfall variations, and thus the need for analysis of rainfall statistics at different locations [1]. There are three conditions to be satisfied in order to produce rainfall:

Saturation conditions must occur such that a thermodynamic state of dew point is satisfied. This is almost exclusively brought about by cooling which accompanies an ascending movement of moist air.

The water content must change from the vapor phase to liquid and/or solid state.

The small water droplets or ice crystals must grow to a precipitable size.

It is common for the first two conditions to occur, but not the third. When the third does not occur, the cloud gradually dissipates because of warming accompanied by evaporation and sublimation.

The three mechanisms for cooling are as follows:

Cyclonic cooling: The nonfrontal type of cyclonic cooling is dependent
   on the convergence and subsequent lifting of air which accompanies
   a low-pressure area. The frontal type is associated with air circula-
   tion that forces moisture-laden air up over a frontal surface. Warm-
   front and low-pressure conditions cause long-duration pelting rains
   and high winds. (Cyclonic cooling is not to be confused with
   hurricanes.)
Convective cooling: This is caused by vertical lifting associated with
   surface heating, i.e., thunderstorms. The resulting precipitation
   is usually of a short duration, but of high intensity.
Orographic cooling: When moist winds blow up a slope, the air will ex-
   pand and cool at the low pressure corresponding to the higher elevation. The direction and velocity of prevailing winds in relation to the
   land slopes involved are important. This phenomenon can be combined with the other mechanisms above.

A major point of interest in rainfall phenomena is storm structure. Figure 4.1 depicts the components of large storm systems which are said to be
synoptic-scale events and which are typically associated with fronts and/or
intensive low-pressure centers.

The components within a synoptic event are typical of other events. For
example, the mesoscale event is typified by a thunderstorm. The mesoscale
size is typically 8 to 50 km. As shown, there may be several mesoscale events
moving within a synoptic-scale event. Within the mesoscale event, there are
convective cells or microscale events. These cells are dynamic, growing and
decimating, and moving with the mesoscale event.

Within a basin a continuous recording rain gage will catch time-varying
amounts of rainfall. However, the above explanations point out that the instantaneous rainfall at any other point will be different and that the average
rainfall over the basin will be different.

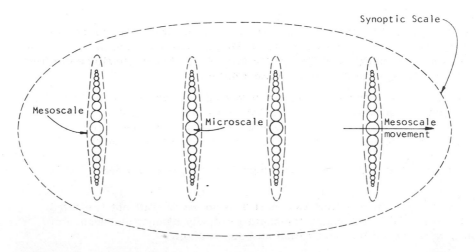

Figure 4.1 Typical instantaneous structure of a synoptic-scale event. (From
Dynamic Hydrology by Peter S. Eagleson. Copyright 1970 by McGraw-Hill
Book Company. Used with the permission of McGraw-Hill Book Company.)

4.2 INFORMATION RESOURCES OF THE UNITED STATES

The U.S. Weather Bureau and recently the National Oceanic and Atmospheric
Administration have prepared many documents which comprise the principal
source of information for rainfall runoff hydrology. The following excerpts
from NOAA Atlas 2 [10] summarize this information.

Historical Review

> The first generalized study of the precipitation-frequency regime for the
> United States was prepared in the early 1930's by David L. Yarnell [2].
> Yarnell's publication contains a series of generalized rainfall maps for
> durations of 5 minutes to 24 hours for return periods of 2 to 100 years.
> Yarnell's study served as a basic source of frequency data for economic
> and engineering design until the middle 1950's. The maps were based on
> data from about 200 first-order Weather Bureau stations equipped with
> recording precipitation gages.

> *Weather Bureau Technical Paper No. 24*, published in two parts [3], was
> prepared for the Corps of Engineers, in connection with its military con-
> struction program. This Technical Paper contained the results of the
> first investigation of precipitation-frequency information for an extensive
> region of the increased hydrologic data network. The results showed
> the importance of the additional data for defining the short-duration
> rainfall-frequency regime in a mountainous region of the western United
> States. In many instances, the differences between the values given in
> Technical Paper No. 24 and those given by Yarnell reach a factor of
> three, with Yarnell's figures generally higher. Results from these two
> studies in the United States were then used to prepare similar reports
> for the coastal regions of North Africa [4] and for several Arctic regions
> [5] where recording-gage data were lacking. These reports were also
> prepared in cooperation with the Corps of Engineers to support its mili-
> tary construction program.

> In 1955, the Weather Bureau and the Soil Conservation Service began
> a cooperative effort to define the depth-area-duration precipitation
> frequency regime in the entire United States. *Weather Bureau Tech-
> nical Paper No. 25* [6], partly a byproduct of previous work done
> for the Corps of Engineers, was the first study published under the
> sponsorship of the Soil Conservation Service; it contains a series of
> precipitation intensity-duration-frequency curves for about 200
> first-order Weather Bureau stations. This was followed by *Weather
> Bureau Technical Paper No. 28* [7] which was an expansion of information
> contained in Technical Paper No. 24 to longer return periods and dura-
> tions. The five parts of *Weather Bureau Technical Paper No. 29* [8], for
> the region east of longitude 90°W., were published next. This Technical
> Paper included seasonal variation on a frequency basis and area-depth
> curves so that the point-frequency values could be transformed to
> areal-frequency values. . . .

> In the next study, *Weather Bureau Technical Paper No. 40* [9], the
> results of previous Weather Bureau investigations of the precipitation-
> frequency regime of the conterminous United States were combined
> into a single publication. Investigations by the Weather Bureau dur-
> ing the 1950's had not covered the region between longitudes 90° and

105°W. Technical Paper No. 40 contained the results of an investigation for this region, and was the first such study of the mid-western
plains region since Yarnell's work of the early 1930's. Topography
was considered only in a general sense in this and earlier studies.

Technical Paper No. 40 was accepted as the standard source for precipitation-frequency information in the United States through 1973 and continues
to be used for the Eastern and Central United States. A sample isopluvial
map of the 49 maps in this document is presented in Fig. 4.2. The maps
cover the 1-, 2-, 5-, 10-, 25-, 50- and 100-year frequencies and 30-min,
1-hr, 1-hr, 3-hr, 6-hr, 12-hr, and 24-hr durations. NOAA Technical Memorandum NWS HYDRO-35 augments Technical Paper No. 40, providing 5- to 60-min
precipitaton values for the Eastern and Central United States [25].
Results presented in Technical Paper No. 40

are most reliable in relatively flat plains. While the averages of point
values over relatively large mountainous regions are reliable, the variations within such regions are not adequately defined. In the largest of
these regions, the western United States, topography plays a significant
role in the incidence and distribution of precipitation. Consequently,
the variations in precipitation-frequency values are actually greater than
portrayed in the region.

Precipitation-Frequency Atlas of the Western United States, NOAA Atlas
2, was published in 1973 [10] to refine information for this region. Each volume of this atlas contains precipitation-frequency maps for 6- and 24- hour
durations for return periods from 2 to 100 years for one of the 11 western
states (west of about 103°W). Also included are methods and nomograms for
estimating values for durations other than 6 and 24 hr. This new series of
maps differs from previous publications through greater attention to the relation between topography and precipitation-frequency values. This relation is
studied objectively through the use of multiple regression screening techniques which develop equations used to assist in interpolating values between
stations in regions of sparse data. The maps were drawn on a scale of
1:1,000,000 and reduced to 1:2,000,000 for publication. Figure 4.3 is an
example of one of the 12 maps for Colorado.
In addition to the maps, each volume includes a historical review of
precipitation-frequency studies, a discussion of the data handling and analysis methods, a section on the use and interpretation of the maps, and a section outlining information pertinent to the precipitation-frequency regime in
the individual state. The section regarding the states includes a discussion
of the importance of snow in the precipitation-frequency analysis and formulas
and nomograms for obtaining values for 1-, 2-, 3-, and 12-hr durations.
Though Technical Paper No. 40 has maps for maximum probable precipitation, much more information has been generated. Some key documents include: Hydrometeorological Report No. 51, *Probable Maximum Precipitation
Estimates, United States East of the 105th Meridian* [26], Hydrometeorological
Report 49, *Probable Maximum Precipitation Estimates, Colorado River and
Great Basin Drainages* [27], and *Design of Small Dams*, 2nd ed. [28]. The
last named publication illustrates the use of probable maximum precipitation information, but the maps therein are based on earlier analysis. Much of the above
information was derived through various cooperative efforts of the Weather
Bureau, National Oceanic and Atmospheric Administration, Bureau of Reclamation, Army Corps of Engineers, and Soil Conservation Service.

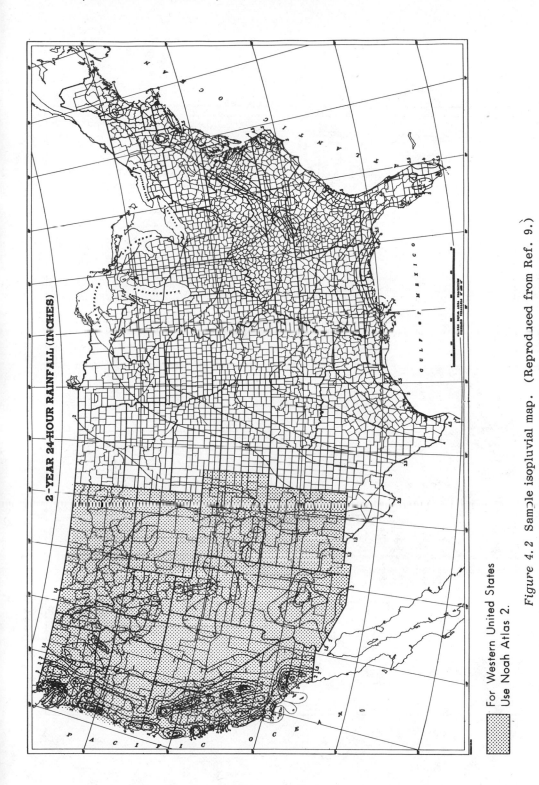

Figure 4.2 Sample isopluvial map. (Reproduced from Ref. 9.)

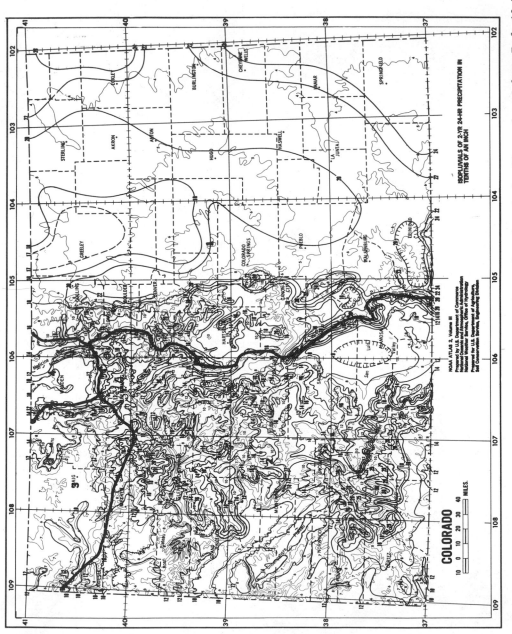

Figure 4.3 Isopluvials of 2-year 24-hr precipitation in tenths of an inch. (Reproduced from Ref. 10.)

Approach Used for Preparation of NOAA Atlas 2

Knowledge about the preparation of rainfall statistics is useful to assure
correct application. Also, it will allow the engineer to recognize unusual or
more severe situations. NOAA Atlas 2 has a narrative explanation of the
approach used therein. It is generally applicable and presents an analysis
of factors to which various rainfall statistics are sensitive. Further excerpts
from this document are presented below.

> The approach used for the Atlas is basically the same as that used for
> Technical Paper No. 40, in which simplified relations between duration
> and return period were used to determine numerous combinations of re-
> turn periods and durations from several generalized key maps. For the
> Atlas, relations were developed between precipitation-frequency values
> and meteorologic and topographic factors at observing sites. These
> were used to aid in interpolating values between stations on the key
> maps.

> The key maps developed were for 2 and 100-year return periods for 6
> and 24-hour durations. The initial map developed was for the 2-year
> return period for the 24-hour duration. This return period was selected
> because values for shorter return periods can be estimated with greater
> reliability than for longer return periods. The 24-hour duration was
> selected because this permitted use of data from both recording and non
> recording gages. Also, because an extensive nonrecording-gage net-
> work was in existence for many years before the recording-gage network
> was established in 1940, the period of record available for 24-hour ob-
> servations is much longer than that for the 6-hour duration. The second
> map developed was for the 100-year return period for the 24-hour dura-
> tion. In the development of this map the advantage of maximum sample
> size and length of record was retained at the expense of some decrease
> in reliability of computed values. The 6-hour maps for the 2 and
> 100-year return periods followed. For the 6-hour duration, the sample
> size was materially smaller in both numbers and length of record because
> only recording-gage data could be used. After these four maps were
> completed, values for intermediate return periods were computed for a
> grid of about 47,000 points, and appropriate maps were prepared.

> In previous studies, such as Technical Paper No. 40, topography was
> considered only in a general sense and the isopluvials were drawn by
> interpolating subjectively between the individual stations. In preparing
> the Atlas for the western states, multiple linear regression equations
> were developed for each of many regions as an aid to estimating the
> precipitation-frequency values at each of about 47,000 grid points.
> These equations related topographic and climatologic factors to the vari-
> ations in the precipitation-frequency values. Isopluvials were smoothed
> subjectively between values in adjoining regions. The subjective
> smoothing was based upon experience in analyzing precipitation-
> frequency maps; the amount of smoothing was rarely greater than the
> standard error of estimate for the equations in the adjoining regions.

Data Sources

The primary data sources used were *Climatological Data for the United
States by Sections* [11] and *Hourly Precipitation Data* [12]. In Califor-

nia, it was possible to increase the data sample 15 to 20 percent by using unpublished data from gages maintained by State, local agencies, private corporations, or individuals [13]. Published data are routinely of high quality because of periodic checks of observing sites and observation techniques and the quality-control procedures used in the publication process. The quality of unpublished data must be checked by a review of the inspection records of the organization maintaining the gage and by a careful screening of the data.

Published and unpublished data from approximately 3300 stations were used in the atlas; 1030 are recording stations and 2292 are nonrecording gages.

Frequency Analysis

There are two methods of selecting data for analysis of extreme values. The first method produces the annual series. This method selects the largest single event that occurred within each year of record. In the annual series, year may be calendar year, water year, or any other consecutive 12-month period. The limiting factor is that one, and only one, piece of datum is accepted for each year. The second method recognizes that large amounts are not calendar bound and that more than one large event may occur in the time unit used as a year. In a partial duration series, the largest N events are used regardless of how many may occur in the same year; the only restriction is that independence of individual events be maintained. The number of events used is at least equal to the number of years of record.

One requirement in the preparation of the Atlas was that the results be expressed in terms of partial-duration frequencies. To avoid the laborious processing of partial-duration data, the annual series data were collected and analyzed and the resulting statistics were transformed to partial-duration statistics.

Conversion Factors between Annual and Partial-Duration Series

Table 4.1 gives the empirical factors used to multiply partial-duration series analysis values to obtain the equivalent annual series analysis values. It is based on a sample of about 200 widely scattered first-order Weather Bureau stations. Only about one-fourth of these stations are in the western United States. The factors used in Table 4.1 were taken

Table 4.1 Empirical Factors for Converting Partial-Duration Series to Annual Series

Return period	Conversion factor
2 years	0.88
5 years	0.96
10 years	0.99

from *Weather Bureau Technical Paper No. 40.* Reciprocals of these factors were used to convert the statistics of the annual series to those of the partial-duration series.

These relations have also been investigated by Langbein [14] and Chow [15] with equivalent results. Basically a ratio of 1.11 between the mean of the partial and annual series results. The means for both series are equivalent to the 2.3-year return period. Tests for samples of from 10 to 50 years of record length indicate that the factors of Table 4.1 are independent of the record length.

Frequency Distribution

The frequency distribution used was the Fisher-Tippett Type I distribution; the fitting procedure was that developed by Gumbel (1958) [16]. Studies by Hershfield and Kohler (1960) [17] and Hershfield (1962) [18] have demonstrated the applicability of this distribution to precipitation extremes. The distribution was fitted by the method of moments. The 2-year value measures the first moment, the central tendency of the distribution. The relation of the 2-year to the 100-year value is a measure of the second moment, the dispersion of the distribution. The 2-year and 100-year precipitation can be used for estimating values for other return periods.

The return-period diagram, Fig. 4.4 taken from *Weather Bureau Technical Paper No. 40*, is based on data from National Weather Service stations having long records. The spacing of the vertical lines on the diagram is partly empirical and partly theoretical. From 1 to 10-year return periods, it is entirely empirical, based on freehand curves drawn through plottings of partial-duration series data. For 20-year and longer return periods, reliance was placed on the Gumbel procedure for fitting annual series data to the Fisher-Tippett Type I distribution. The transition was smoothed subjectively between the 10 and 20-year return periods. If precipitation values for return periods between 2 and 100 years are desired, it is necessary to obtain the 2 and 100-year values from this series of generalized precipitation-frequency maps. These values are then plotted on the appropriate verticals and connected with a straight line. The precipitation values for the intermediate return periods are determined by reading values where the straight line intersects the appropriate verticals. If the rainfall values are then converted to the annual series by applying the factors of Table 4.1 and plotted on either Gumbel or log-normal graph paper, the points will very nearly approximate a straight line.

Development of Relations for Interpolating Precipitation-
Frequency Values for the Atlas

The adequacy of the basic data network for determining precipitation-frequency values varies from place to place within the western United States. The greatest station density occurs along the Pacific coast west of the Cascade and Sierra Nevada Ranges. The lowest densities are in the intermountain plateau—between the Cascade-Sierra Nevada ranges and the Continental Divide—particularly in Nevada and in the Salmon River Mountains of Idaho. Even within particular regions, the stations are not evenly distributed. Most of the stations

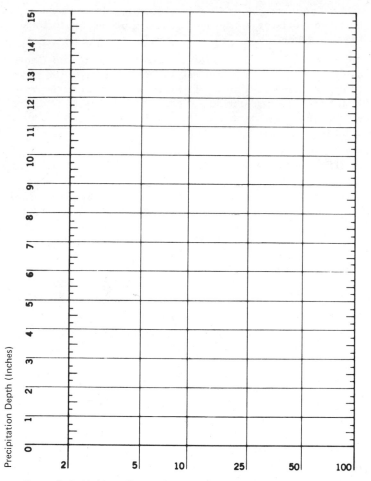

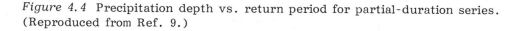

Figure 4.4 Precipitation depth vs. return period for partial-duration series. (Reproduced from Ref. 9.)

are located in the coastal plains, the river valleys, the western portion of the Great Plains, and the lower foothills of the mountains. Relatively few stations are located on steep slopes or on crests of mountains, in sparsely populated areas, or in areas where access is difficult.

It is desirable, therefore, to develop relations that can be used in interpolating precipitation-frequency values between stations in regions where data are relatively scarce. A preferred method is to relate variations in precipitation-frequency directly to variations in topographic factors; this was done when an adequate relation could be developed. The primary advantage of this procedure is that topographic factors can be determined at any point in a region. Topographic maps can be prepared from aerial photographs or surveys, or by other methods that do not require observations taken at a fixed point over a period of time. Among topographic factors frequently considered are: (1) elevation of the station, either the actual elevation or some effective elevation (an average elevation determined along a circle of a given radius around the station); (2) slope of the terrain near the station, both in the small and large scales; (3) distances from both major and minor barriers; (4) distances and directions from moisture sources; and (5) roughness of the terrain in the vicinity of the station.

It would have been desirable to develop a single equation, utilizing physiographic factors, to interpolate between locations with short-duration precipitation-frequency values for the western United States. Such an equation could not be developed, so relations for interpolating the precipitation-frequency values were developed for each of several smaller regions considered to be meteorologically homogeneous. The extent of each region was determined from consideration of the weather situations that could be expected to produce large precipitation amounts. Among the questions asked and answered were: What is the source and from what direction does moisture for major storms come and are there major orographic barriers that influence the precipitation process? Figure 4.5 shows some of the principal paths of moisture inflow for the western United States and the major orographic barriers to such inflow.

The regions selected for their homogeneity normally are river basins or combinations of river basins. The river basins selected were usually bounded by major orographic barriers that significantly influence the precipitation regime.

After the geographic regions were selected, various topographic factors that could cause variation of precipitation-frequency values within limited regions such as slope, elevation, roughness, and orientation were examined. Individual precipitation-frequency values and exposures around the stations were examined to gain insight into topographic factors that could be important.

It was not possible to develop such relations for all regions. Hence, it also was necessary to develop relations that included climatological or meteorological factors. The factors selected for use must be available at locations where precipitation data for durations of between 1 and 24 hours are not available. Otherwise, they would not provide additional information needed for use in interpolating between locations with frequency values. An example of such a factor is normal annual precipitation. In the construction of such a map, data from snow courses, adjusted short records, and storage gages that give weekly, seasonal, or annual accumulations of precipitation can be used. Such records do not yield the short-duration precipitation amounts necessary for this study. Thus, normal annual precipitation data, particularly because it provides greater areal coverage in mountainous regions, might be of definite use in developing the patterns of the precipitation-frequency maps. Several other meteorologic factors can be used in combination with normal annual precipitation data and topographic factors to interpolate short-duration precipitation-frequency values at intermediate points. Examples of such factors are: (1) number of thunderstorm days (2) number of days or hours with precipitation above a threshold value, (3) percentage

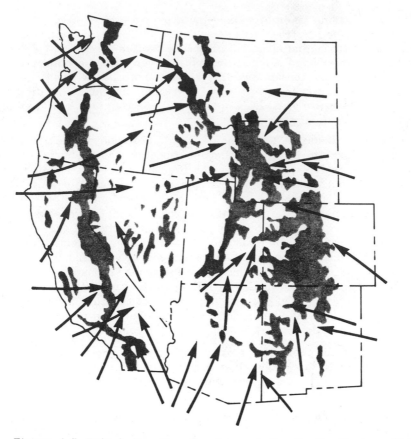

Figure 4.5 Principal paths of moisture inflow in the Western United States for storms producing large precipitation amounts. Toned areas are major orographic barriers. (Reproduced from Ref. 10.)

frequencies of various wind directions and speeds, and (4) percentage frequencies of class intervals of relative humidity. Since these factors can be obtained only where there are recording meteorological gages or where there are observers to record the data they do not supplement the available short-duration precipitation-frequency values by providing data at additional sites.

Finally, various climatological and meteorological factors that could be indexes of variation of the precipitation-frequency values were considered. The procedure used for developing interpolating equations was a multiple-regression screening technique. This process was done by computer using a least-squares technique. The number of variables screened for the various relations ranged between 60 and 100. This does not mean that 60 or more completely different factors could be identified. For example, several factors might involve different measures of slope. Moreover, these measures of slope might be over different distances or have different orientations. In each instance, the practice was to permit the computer to select the most critical of the various measures of each factor.

The computer program selected the single variable most highly correlated with the precipitation-frequency value under investigation. The next step was to select the variable that, combined with the variable already selected, would explain the greatest variation in the precipitation-frequency values. The third, fourth, fifth, and further variables were selected in a similar manner. The program continued to select variables until the variance explained by an additional variable was less than some preselected amount or until a fixed number of variables was selected. Final equations did not contain more than five independent variables.

Relations for Interpolating between 24-hour
Precipitation-Frequency Data Points

The equations developed for interpolating between locations with 2-year 24-hour precipitation values in regions of sparse data were not all equally reliable. On the average, the 28 equations developed for estimating the 2-year 24-hour precipitation values at intermediate points in western United States explained about 70 percent of the variance. The standard error of estimate averaged about 13 percent of the average station value for 2-year 24-hour precipitation.

Table 4.2 shows the factors, grouped in general categories, found most useful in depicting variations in the 2-year 24-hour precipitation values for the western United States. The first and second columns show the number and percent of equations in which each factor was used. The total for the second column is larger than 100 percent because several factors were used in the equations developed for each region. The third column shows the total number of times each factor was used, and the fourth what percentage each factor used was of the total number of factors. For example, of the 89 different factors used in the 28 equations, 37 were some measure of slope; the use of the slope factor represents 42 percent of the total number of factors used.

The single most important factor considered was slope, a topographic factor. Measurement of slope varied from region to region. In some regions, slope was measured directly by dividing the difference in height between two points by the distance between the points.

The second most important topographic factor was found to be the barrier to moist airflow; this factor is actually a combination of meteorology and topography. In selecting a barrier, the first consideration was the direction of moist air inflow. The barrier had to be normal, or nearly normal, to this direction. The barrier range, or ranges, had to be sufficiently massive to cause a significant disruption in the airflow. Barriers of limited lateral extent that would permit air to flow around as easily as over were not considered.

The distance to the principal moisture source, a combination of topographic and meteorologic influences, was another important factor. In northeastern New Mexico, central Colorado, and southeastern Wyoming, examination of a topographic map and consideration of the moist air inflow in storms that produced large precipitation amounts (Fig. 4.5) made it evident that the general moist airflow was from the Gulf of Mexico. Distance to moisture was therefore measured in that direction.

Table 4.2 Factors Most Useful in Relations for Interstation Interpolation for 2-Year 24-hr Precipitation Values

Factor (by category)	No. of equations using factor	Percent of equations using factor	No. of times each factor was used	Percent of total no. of times each factor was used
Slope	18	64	37	42
Normal annual precipitation	15	54	15	17
Barrier to airflow	10	36	11	12
Elevation	10	36	10	11
Distance to moisture	9	32	9	10
Location (latitude or longitude)	4	14	5	6
Roughness	2	7	2	2

Source: NOAA Atlas 2 [10].

Table 4.3 shows the factors found most useful for interpolating variations in the 100-year 24-hour precipitation values in sparse-data areas of the western United States. This table is in the same format as Table 4.2. The definitions of the variables—slope, distance to moisture, elevation, etc.—are the same as those for Table 4.2. In the equations, the 2-year 24-hour precipitation values were used in interpolation.

Relations for Estimating the 6-hour Precipitation-Frequency Values
Data from both recording and nonrecording gages can be incorporated in equations for estimating precipitation-frequency values for the 24-hour duration. For durations of less than 24 hours, only data from recording gages can be used. This frequently reduces the number of data points within a particular region by one-half or more. The effect of topography on precipitation-frequency values decreases as the duration decreases. Thus, there is less variability in the precipitation-frequency values for the 6-hour duration. For these reasons, larger regions are used to develop interpolation equations for 6-hour duration maps.

The factors used most frequently in the equations for estimating the 2-year 6-hour precipitation values are listed in Table 4.4, those for the 100-year 6-hour precipitation values are given in Table 4.5. The format and definitions of variables of Tables 4.4 and 4.5 are the same as those of Table 4.2.

Table 4.3 Factors Most Useful in Relations for Interstation Interpolation for 100-Year 24-hr Precipitation Values

Factor (by category)	No. of equations using factor	Percent of equations using factor	No. of times each factor was used	Percent of total no. of times each factor was used
2-year 24-hr precipitation	27	77	27	29
Slope	26	74	26	28
Elevation	20	57	20	22
Distance to moisture	6	17	6	7
Location (latitude or longitude)	5	14	6	7
Normal annual precipitation	4	11	4	4
Barrier to inflow	3	6	2	2
Roughness	1	3	1	1

Source: NOAA Atlas 2 [10].

Table 4.4 Factors Most Useful in Relations for Interstation Interpolation of 2-Year 6-hr Precipitation Values

Factor (by category)	No. of equations using factor	Percent of equations using factor	No. of times each factor was used	Percent of total no. of times each factor was used
Slope	4	40	10	38
2-year 24-hr precipitation	7	70	7	27
Location (latitude or longitude)	4	40	4	15
Elevation	3	30	3	12
Barrier to airflow	1	10	1	4
Distance to moisture	1	10	1	4

Source: NOAA Atlas 2 [10].

Table 4.5 Factors Most Useful in Relations for Interstation Interpolation for 100-Year 6-hr Precipitation Values

Factor (by category)	No. of equations using factor	Percent of equations using factor	No. of times each factor was used	Percent of total no. of times each factor was used
2-year 24-hr precipitation	5	55	5	23
100-year 24-hr precipitation	4	36	4	19
Elevation	4	36	4	19
Slope	4	36	4	19
2-year 24-hr precipitation	1	9	1	5
Normal annual precipitation	1	9	1	5
Distance to moisture	1	9	1	5
Location	1	9	1	5

Source: NOAA Atlas 2 [10].

Intermediate Maps

The 47,000-point grid described earlier was also used in the analysis of the isopluvial patterns of the eight intermediate maps. These maps—for 5, 10, 25, and 50-year return periods for 6 and 24-hour durations—were prepared primarily for the convenience of the user, because it is technically sufficient to provide two points of the frequency curve for a particular duration and to describe the method of interpolation. Four values, one from each of the four key maps, were read for each grid point. These four values were used in a computer program based on the return-period diagram (Fig. 4.4) to compute values for eight additional maps. The key maps were used as underlays to maintain the basic isopluvial pattern on all maps.

Procedures for Estimating Values for Durations Other Than 6 and 24 hours

The isopluvial maps in the Atlas are for 6 and 24-hour durations. For many hydrologic purposes, values for other durations are necessary. Such values can be estimated using the 6 and 24-hour maps and the empirical methods outlined in the Atlas as briefly described as follows. The procedures detailed below for obtaining 1, 2, and 3-hour estimates were developed specifically for the Atlas. The procedures for obtaining estimates for less than 1-hr duration and for 12-hr duration were adopted from *Weather Bureau Technical Paper No. 40* [9] only after investigation demonstrated their applicability to data from the area covered by the Atlas.

Procedures for Estimating 1-hr (60-min)
Precipitation-Frequency Values

Multiple-regression screening techniques were used to develop equations for estimating 1-hour duration values. Factors considered in the screening process were restricted to those that could be determined easily from the maps of this Atlas or from generally available topographic maps.

Example equations to provide estimates for the 1-hr duration for 2- and 100-year return periods are shown in Table 4.6 for Colorado regions. For return periods between 2 and 100 years, a nomograph similar to the one in Fig. 4-6a can be utilized.

Estimates for 2- and 3-hr (120- and 180-min)
Precipitation-Frequency Values

Similar regression techniques developed equations which give the 2- and 3-hr precipitation values based upon the 1- and 6-hr values. These equations are also presented in nomograph form (see Fig. 4.6b for Colorado Region 1

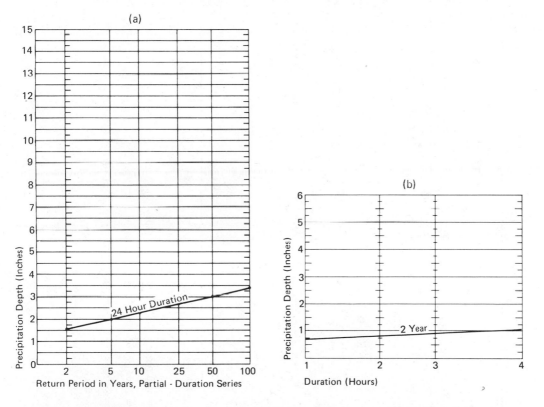

Figure 4.6 Illustration of use of precipitation-frequency diagrams using values from precipitation-frequency maps and relations at 106°00'W, 39°00'N. (Reproduced from Ref. 9.)

Table 4.6 Equations for Estimating 1-Hr Values in Colorado with Statistical Parameters for Each Equations

Region of applicability	Equation	Corr. coeff.	No. of stations	Mean of computed stn. values (in.)	Standard error of estimate (in.)
South Platte, Republican, Arkansas, and Cimarron River basins	$Y_2 = 0.218 + 0.709[(X_1)\ (X_1/X_2)]$	0.94	75	1.01	0.074
	$Y_{100} = 1.897 + 0.439[(X_3)\ (X_3/X_4)] - 0.008Z$	0.84	75	2.68	0.317
San Juan, Upper Rio Grande, Upper Colorado, and Gunnison River basins and Green River basin below confluence with the Yampa River	$Y_2 = -0.011 + 0.042[(X_1)\ (X_1/X_2)]$	0.95	86	0.72	0.085
	$Y_{100} = 0.494 + 0.755[(X_3)\ (X_3/X_4)]$	0.90	85	1.96	0.290
Yampa and Green River basins above confluence of Green and Yampa Rivers	$Y_2 = 0.019 + 0.711[(X_1)\ (X_1/X_2)] + 0.001Z$	0.82	98	0.40	0.031
	$Y_{100} = 0.338 + 0.670[(X_3)\ (X_3/X_4)] + 0.001Z$	0.80	79	1.04	0.141
North Platte	$Y_2 = 0.028 + 0.890[(X_1)\ (X_1/X_2)]$	0.93	90	0.60	0.062
	$Y_{100} = 0.671 + 0.757[(X_3)\ (X_3/X_4)] - 0.003Z$	0.91	88	1.71	0.236

List of Variables:

Y_2 = 2-year, 1-hr estimated value

Y_{100} = 100-year, 1-hr estimated value

X_1 = 2-year, 6-hr value from precipitation-frequency maps

X_2 = 2-year, 24-hr value from precipitation-frequency maps

X_3 = 100-year, 6-hr value from precipitation-frequency maps

X_4 = 100-year, 24-hr value from precipitation-frequency maps

Z = point elevation in hundreds of feet

Source: NOAA Atlas 2, Vol. III [10].

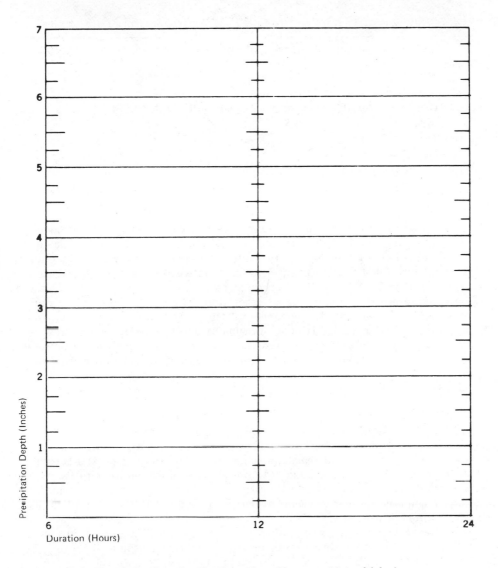

Figure 4.7 Precipitation depth-duration diagram (6 to 24 hr).

example). These nomograms are independent of return period. They were developed using data from the same regions used to develop the 1-hr equations.

Estimates for 12-hr (720-min) Precipitation-
Frequency Values
Nomographs are also available to obtain estimates for the 12-hr duration based upon the 6- and 24-hr maps. Figure 4.7 is an example of this type of nomograph.

Estimates for Less Than 1 hour

To obtain estimates for durations of less than 1 hr, apply the values in Table 4.7 to the 1-hr value for the return period of interest.

Table 4.7 Adjustment Factors to Obtain N-min Estimates from 1-hr Values

Duration (min.):	5	10	15	30
Ratio to 1 hr:	0.29	0.45	0.57	0.79

Adapted from U.S. Weather Bureau Technical Paper No. 40 [9].

Reliability of Results

The locations of the stations used in the analyses are shown in the Atlas and Technical Paper No. 40.

> This geographic network of stations does not reveal with complete accuracy the very detailed structure of the isopluvial patterns in the mountainous regions of the West. The multiple regression equations discussed earlier were used to help in interpolation between values computed for these stations. The standard error of estimate for these relations should be considered when using the precipitation-frequency values shown on the maps. In general, the accuracy of the estimates obtained from the maps of the Atlas varies from a minimum of about 10 percent for the shorter return periods in relatively nonorographic regions to 20 percent for the longer return periods in the more rugged orographic regions.

Illustration of Use of Precipitation-Frequency Maps, Diagrams, and Equations

To illustrate the use of these maps, values were read from isopluvial maps for Colorado at 39°00'N and 106°00'W. These values are shown in italic type in Table 4.8. The values read from the maps should be plotted on the return-period diagram of Fig. 4.4 because "(1) not all points are as easy to locate on a series of maps as are latitude-longitude intersections, (2) there may be some slight registration differences in printing, and (3) precise interpolation between isolines is difficult."

The 24-hr values shown in Table 4.8 were plotted on Fig. 4.6a, and a line of best fit has been drawn subjectively. On this nomogram the line fits the data rather well. Had any points deviated noticeably from the line, the value would have been reread from the line and the new value substituted in Table 4.8 and adopted in preference to the original readings.

The 2- and 100-year 1-hr values for the point were computed from the equations applicable to Region 1 in Table 4.6, since the point is east of the Continental Divide. The 2-year 1-hr value is estimated at 0.71 in. (2-year 6- and 12-hr values from Table 4.8); the estimated 100-year 1-hr value is 1.86 in. (100-year 6- and 24-hr values from Table 4.8 and elevation of 9500 ft). By plotting these 1-hr values on Fig. 4.4 and connecting them with a straight line, one can obtain estimates for return periods of 5, 10, 25, and 50 years.

The 2- and 3-hr values can be estimated by using the nomogram for the region as shown in Fig. 4.6b. The 1- and 6-hr values for the desired return

Table 4.8 Precipitation Data for Depth-Frequency Atlas Computation Point
106°00'W, 39°00'N

Return period	1 hr	2 hr	3 hr	6 hr	24 hr
2 years	0.71	0.83	0.91	1.05	1.58
5 years				1.38	1.99
10 years				1.59	2.27
25 years				1.90	2.65
50 years	1.86			2.19	2.95
100 years				2.39	3.35

period are obtained as above. Read the estimates for 2 or 3 hrs at the intersections of the connecting line and the 2- and 3-hr vertical lines. The 2-year
2-hr (0.83-in.) and 2-year 3-hr (0.91-in.) values are in italics in Table 4.8.
The 12-hr value can be estimated by using Fig. 4.7.

4.3 REGIONAL EXAMPLE

Many regional agencies have developed detailed isopluvial mapping and other
rainfall information. The following paragraphs describe the approach in the
Denver, Colorado region [24].

Denver Regional Approach

The method of analysis used in this effort was similar to that employed by the
U.S. Weather Bureau in several of their generalized rainfall-frequency studies
[7-9]. Simplified duration and return-period relationships were developed for
use with several key rainfall-frequency maps so that any combination of return period from 1 to 100 years and durations from 5 min to 24 hr can be
determined. The rainfall-frequency values and isohyetal maps were based on
point rainfall data from 17 recording stations, 24 nonrecording stations, and
the Stapleton International Airport Weather Bureau Station.
    Examples are given to explain the use of the isohyetal maps and the
graphs. Nine isohyetal maps are presented (Figs. 4.8 through 4.16). These
maps show the depth of rain to be expected for various frequencies with various return periods for the entire six-county area, from Palmer Lake to Longs
Peak, and from Deer Trail to the Continental Divide. Blank "Rainfall
Depth-Duration-Frequency Graphs, Denver Region" are presented for use on
specific problems (Fig. 4.17).

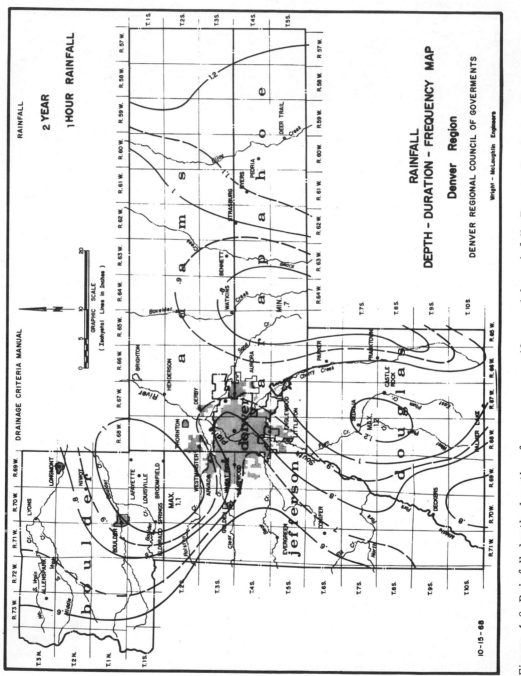

Figure 4.8 Rainfall depth-duration-frequency map (2-year 1-hr rainfall, Denver Region). (Reproduced from Ref. 24.)

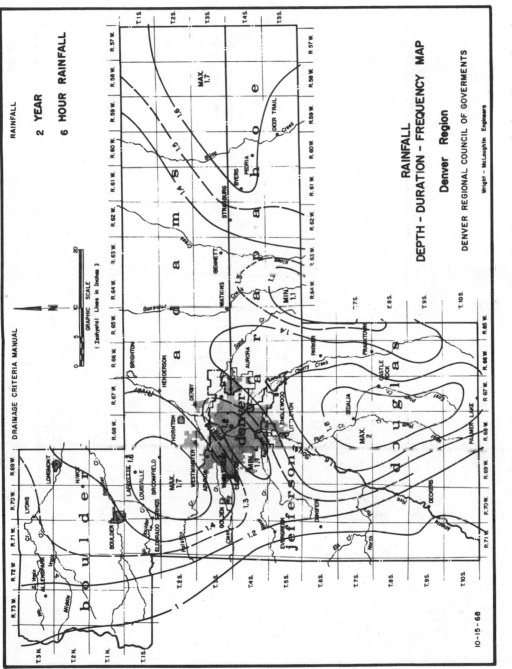

Figure 4.9 Rainfall depth-duration-frequency map (2-year 6-hr rainfall, Denver Region). (Reproduced from Ref. 24.)

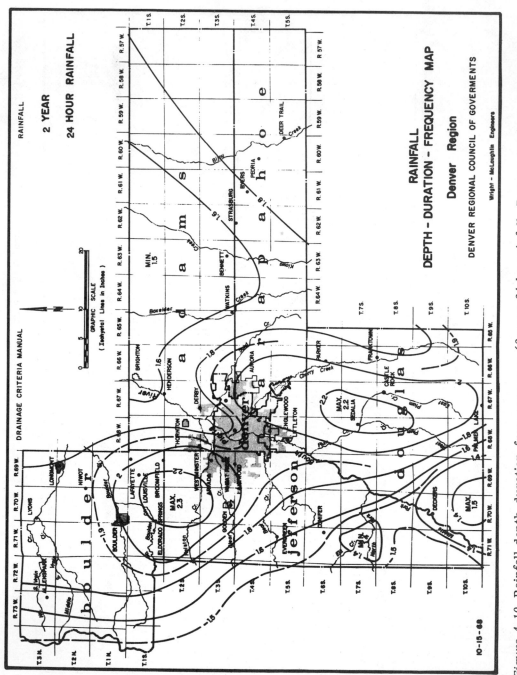

Figure 4.10 Rainfall depth-duration-frequency map (2-year 24-hr rainfall, Denver Region). (Reproduced from Ref. 24.)

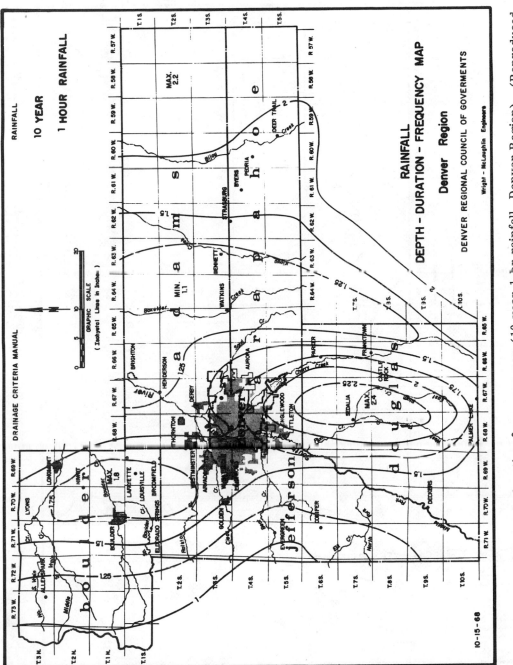

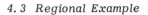

Figure 4.11 Rainfall depth-duration-frequency map (10-year 1-hr rainfall, Denver Region). (Reproduced from Ref. 24.)

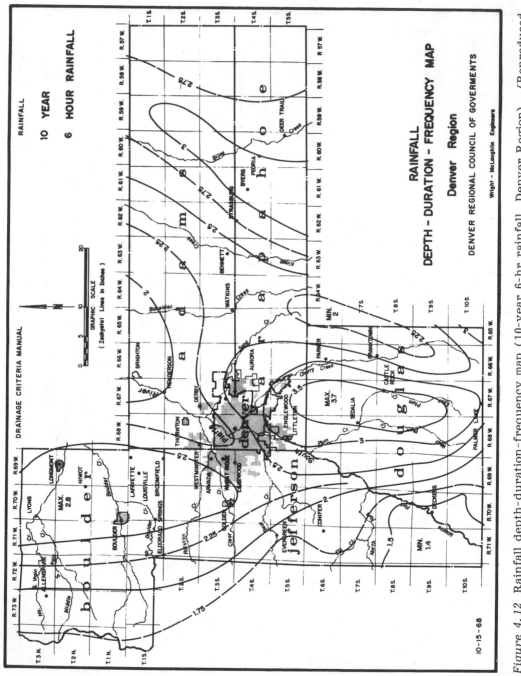

Figure 4.12 Rainfall depth-duration-frequency map (10-year 6-hr rainfall, Denver Region). (Reproduced from Ref. 24.)

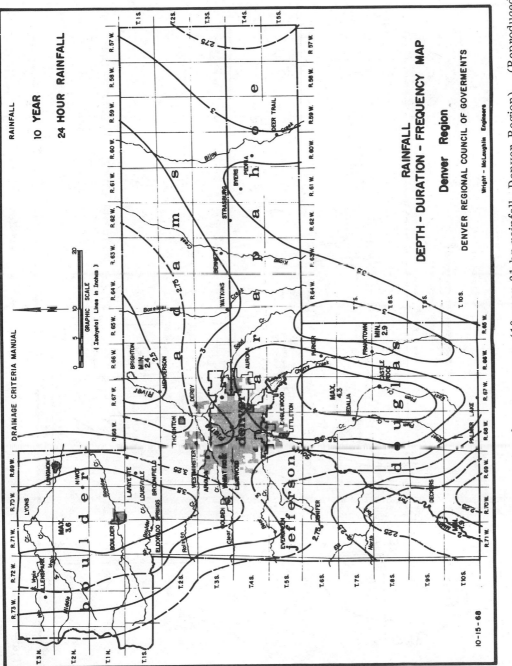

Figure 4.13 Rainfall depth-duration-frequency map (10-year 24-hr rainfall, Denver Region). (Reproduced from Ref. 24.)

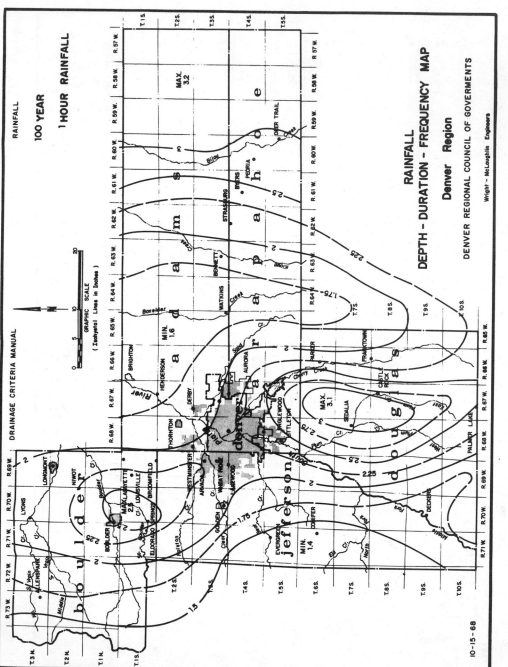

Figure 4.14 Rainfall depth-duration-frequency map (100-year 1-hr rainfall, Denver Region). (Reproduced from Ref. 24.)

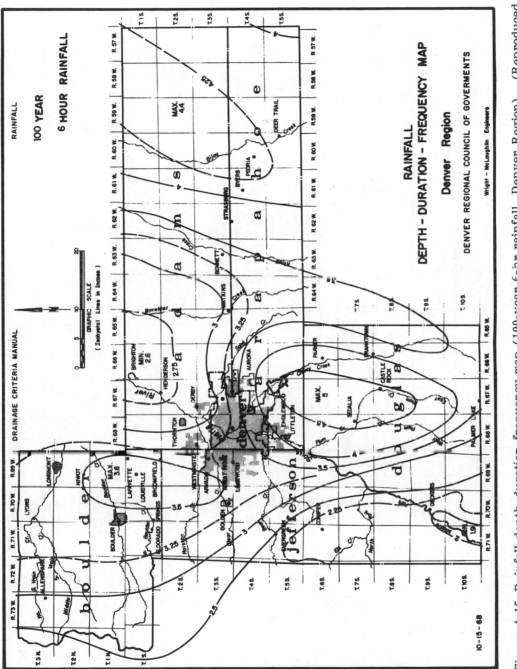

Figure 4.15 Rainfall depth-duration-frequency map (100-year 6-hr rainfall, Denver Region). (Reproduced from Ref. 24.)

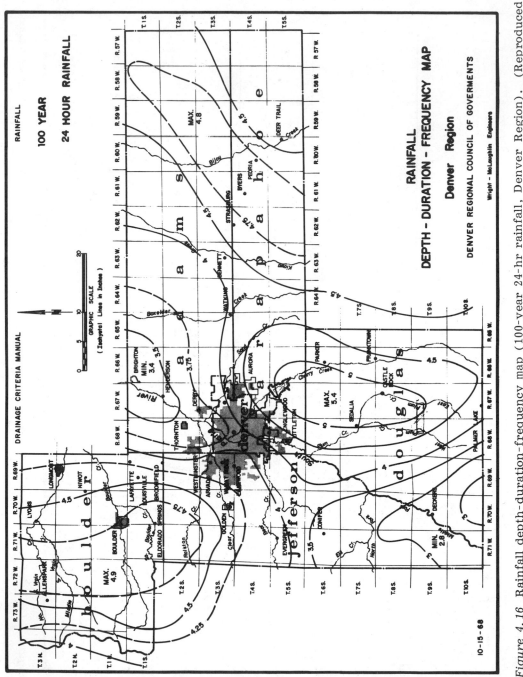

Figure 4.16 Rainfall depth-duration-frequency map (100-year 24-hr rainfall, Denver Region). (Reproduced from Ref. 24.)

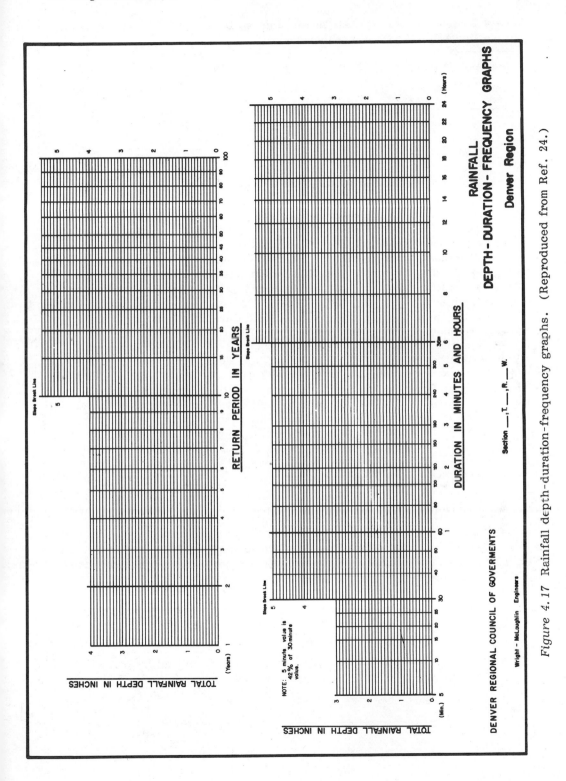

Figure 4.17 Rainfall depth-duration-frequency graphs. (Reproduced from Ref. 24.)

Table 4.9 Relationship between 30-min Rainfall Depth and Shorter-Duration Rainfall Depth for the Same Frequency

Duration D (min):	5	10	15	20	25
Ratio D/30 min:	0.42	0.63	0.75	0.84	0.92

A generalized duration relationship was developed with which the rainfall depth for a selected return period can be determined for any duration from 5 min to 24 hr. Two relationships were considered, one for durations from 30 min to 6 hr and the second for durations from 6 to 24 hr. These are shown in the lower graphs in Figs. 4.18a and b. The data for Denver Weather Bureau Airport provided the primary control in the construction of the 30-min to 6-hr figure. Consideration was given to the rainfall intensity-duration-frequency curves for Denver presented in Weather Bureau Technical Paper No. 25 [6].

The generalized diagram was extended to durations less than 30 min. Average ratios of 5-, 10-, and 15-min rainfall to the 30-min rainfall were obtained from the Denver Weather Bureau Airport data and information contained in Weather Bureau Technical Report No. 25 [6] for Denver. The ratios were nearly identical for each duration for return periods from 2 to 100 years. The values for durations of 5 to 30 min can be obtained from the "Depth-Duration-Frequency Graphs" by multiplying the 30-min value by 0.42 and plotting this new value at the 5-min duration, then drawing a straight line from the 5-min to the 30-min value and reading the precipitation for any duration from 5 to 30 min from the graph. Ratios for 5 through 25 min are shown in Table 4.9. These ratios compare favorably with those presented in Weather Bureau Technical Paper No. 40 [9].

Examples of Obtaining Rainfall-Frequency Data

Example 1: Depth-Duration
The 25-year rainfall intensities are required for Section 6, Township 2 South, Range 67 West. The following data are obtained from the rainfall-frequency charts (in inches):

Return period	1 hr	6 hr	24 hr
10 years	1.35	2.10	2.60
100 years	1.80	2.85	3.50

This information is plotted on the depth vs. return period graphs (Fig. 4.18a) and a straightedge laid along the values, and the intersection with the 25-year return period read to yield:

Return period	1 hr	6 hr	24 hr
25 years	1.50	2.39	2.90

These data are then plotted on the depth vs. duration graph to obtain supplemental rainfall values for durations from 30 min to 24 hr. Values obtained for additional duration are:

Return period	30 min	2 hr	3 hr	12 hr
25 years	1.20	1.80	2.00	2.65

If the rainfalls for durations from 5 to 30 min are required, they can be obtained by multiplying the 25-year 30-min value, 1.15 in., by 0.42 and plotting that value at the 50-min duration point on the depth-duration graph and a straight line drawn between the 5- and 30-min values. The following values are then read:

Return period	5 min	10 min	15 min	20 min	25 min
25 years	0.50	0.76	0.90	1.01	1.10

The 25-year rainfall values, as indicated above, for the various durations can then be plotted on linear paper and a smooth curve drawn through the points to obtain a depth-duration curve. They will plot also as nearly straight lines on logarithmic (log-log) paper with a marked break at 30 min and a slight break at 6 hr. Either of the above plots will yield sufficient rainfall values for drainage design.

Example 2: Development of Intensity-Duration Curves
A series of intensity-duration curves can be developed for the 2-, 5-, 10-, 25-, 50-, and 100-year frequencies to be used in conjunction with the Rational Method formula for Section 31, Township 2 North, Range 70 West. From the rainfall-frequency charts, the following (in inches) can be obtained:

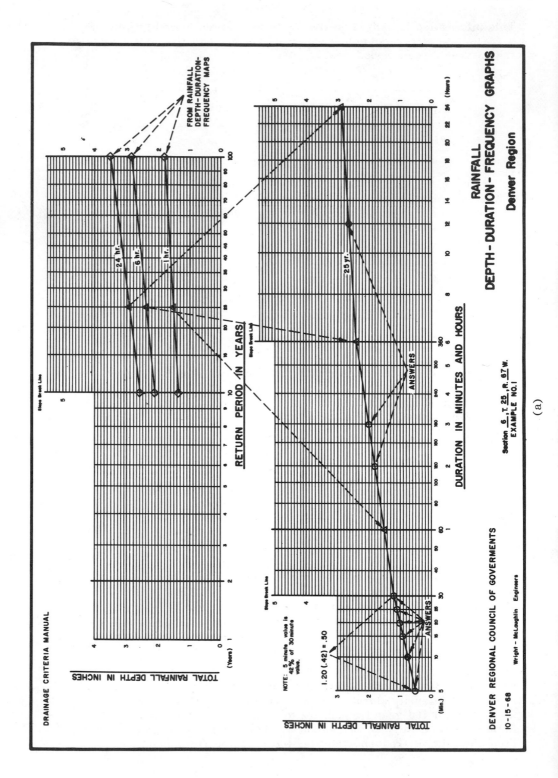

(a)

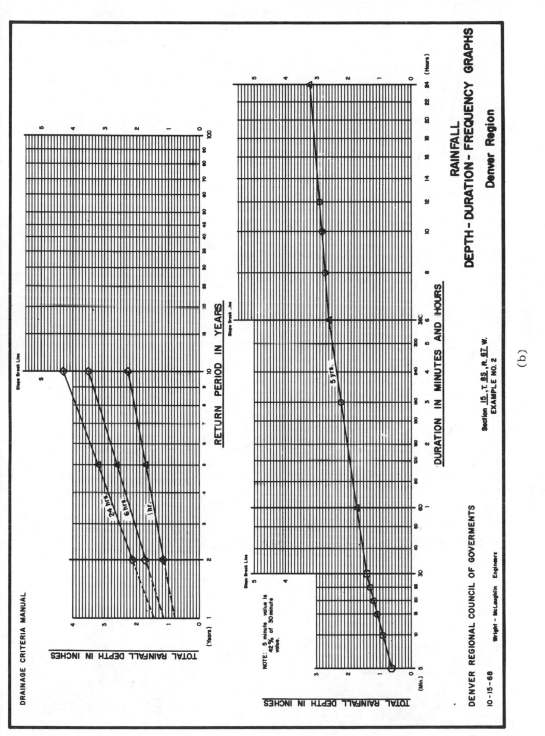

Figure 4.18 Rainfall depth-duration-frequency graphs. (Reproduced from Ref. 24.)

(b)

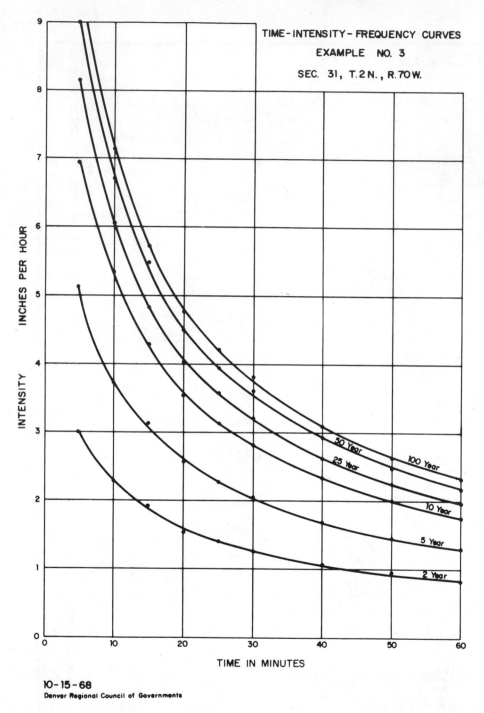

Figure 4.19 Time-intensity-frequency curves. (Reproduced from Ref. 24.)

Return period	1 hr	6 hr	24 hr
2 years	0.83	0.36	1.89
10 years	1.75	2.70	3.60
100 years	2.30	3.40	4.65

After plotting these values on the depth vs. return period graph, various depths of rainfall for the 1-, 6-, and 24-hr durations for the 2-, 5-, 10-, 25-, 50-, and 100-year return periods can be obtained. These values can be plotted on the depth vs. duration graph and straight lines can be drawn on Fig. 4.18b for each frequency. Then the various depths of the various frequencies for several durations can be obtained. These depth-duration values can be converted to intensity values in inches per hour. These data can then be plotted on a time-intensity-frequency graph (see Fig. 4.19). These curves can be used directly for obtaining intensity values for the Rational Method.

4.4 RESOURCES OF OTHER COUNTRIES: AN AUSTRALIAN EXAMPLE

Similar approaches have been taken to produce statistical rainfall information in other countries. An approach used in Australia is presented as an example.

Point Rainfall Intensity-Frequency-Duration Pattern Data

The Australian Department of Science, Bureau of Meteorology, has prepared a document entitled *Point Rainfall Intensity-Frequency-Duration Data, Capital Cities*, Bulletin No. 49, 1974 [19]. This document provides rainfall information needed for drainage analysis of "small-scale drainage schemes in the metropolitan areas of the capital cities."

Additionally, the Bureau of Meteorology has polynomial equations representing rainfall curves. These equations describe intensity-frequency-duration curves over a period of 6 min to 72 hr. Methods to utilize these equations are given for various-sized drainage basins. Rainfall diagrams such as Figs. 4.20 and 4.21 are presented for quick reference and approximations.

Selected material from Bulletin No. 49 is presented as follows:

In the analysis of (rainfall) data the following assumptions were made:

the primary rainfalls for each duration (highest recording in the season or a calendar year) constitute a statistically independent series;

the logarithms of the primary falls have a normal or Gaussian distribution;

the rainfalls estimated from the partial duration series (the series containing the highest falls regardless of year of occurrence) for return periods of 2, 5, and 10 years are 13, 5 and 1 percent higher respectively than similar estimates made from the primary [annual] series; and

the once in one year rainfall can be estimated from the formula:

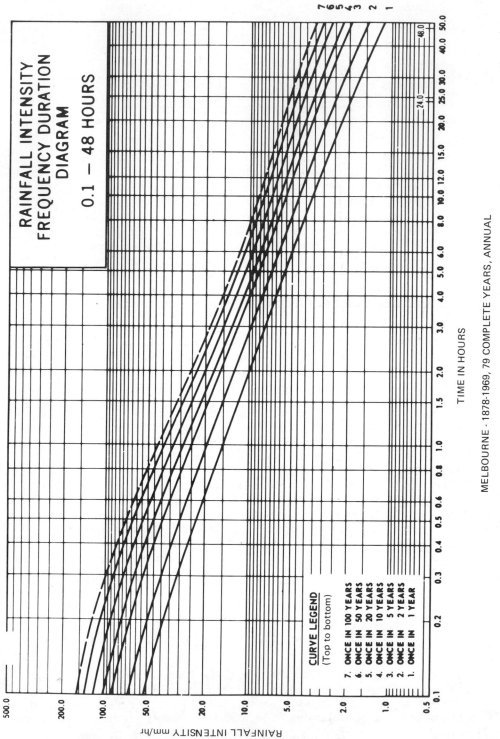

TIME IN HOURS

MELBOURNE - 1878-1969, 79 COMPLETE YEARS, ANNUAL

Figure 4.20 Rainfall intensity frequency duration diagram (0.1 to 48 hr). (Reproduced with permission from Ref. 19.)

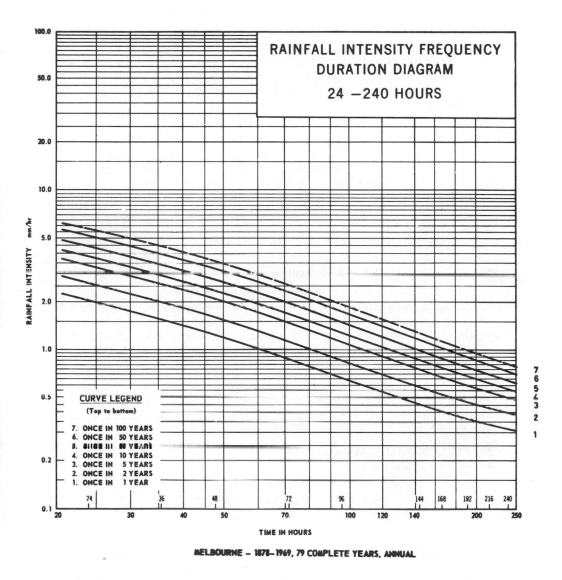

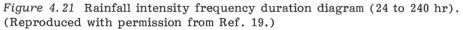

Figure 4.21 Rainfall intensity frequency duration diagram (24 to 240 hr). (Reproduced with permission from Ref. 19.)

$$R = antilog\ (M)/(1 + 0.82S)$$

where R = once in one year rainfall
     M = mean of logarithms of primary rainfalls
     S = standard deviation of logarithms of primary rainfalls

The relationships between the estimates for return periods of 2, 5, and 10 years based on the partial duration series and those based on the primary (annual) series are taken from experience in the United States of America [9]. An examination of data for a limited number of Australian stations indicates that these relationships hold under Australian conditions.

In this study estimates of rainfall for return period of 10 years or less have been based on the primary series rather than the partial duration series for three main reasons. All estimates are uniquely defined by two parameters, namely the mean and the standard deviation of the logarithms of the primary falls. This enables a considerable saving to be made in the computer effort required to produce the desired tabulations and graphs. Also, greater consistency between estimates for various return periods is obtained as individual estimates are not entirely dependent upon a small number of observed values. Finally, each element of the primary series, being, by definition, the greatest fall in a calendar year or a season can be assumed statistically independent of all other elements.

Elements of the partial duration series have no similar time restrictions and, as a result, two elements may be separated by only a short period of time. As there is strong evidence of serial correlation of rainfalls the statistical independence of elements of the partial duration series cannot be assumed.

The assumption of log-normality has been tested on a large number of Australian stations for durations ranging from 6 minutes to 10 days. This distribution has been found to fit the observation at least as well as the others tested and is more easily handled mathematically.

The method of analysis used and the assumptions made in the preparation of data for Bulletin No. 49 are discussed in detail by Pierrehumbert [19].*

Local Adjustments

Locations near capital cities will not have the same rainfall statistics. However, various adjustment factors that allow use of the capital cities' data for the nearby areas are provided.
     Figure 4.22 is an example of an adjustment tool for the Melbourne area.

---

*Reprinted with permission from Ref. 19.

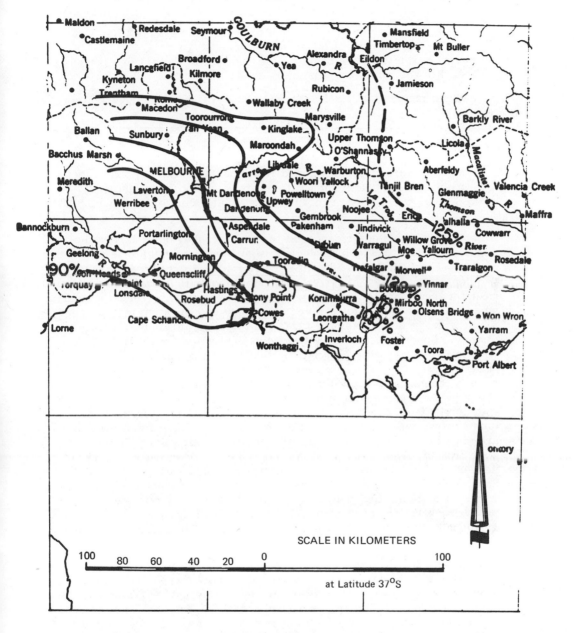

Figure 4.22 Map of adjustment factors for rainfall intensity as related to Melbourne gage.

4.5 SPECIAL CONSIDERATIONS

Depending upon the type of rainfall statistical analysis there may be need for adjustment of rainfall data to take into consideration the following phenomena:

> Rainfall can increase or decrease due to orographic effects when comparing locations with different elevations.
> The average rainfall for a basin larger than 3 km^2 (1 mi^2) is generally less than the rainfall at a point reading.

Adjustment factors for these phenomena are presented in the following paragraphs.

Areal Analysis

The U.S. Weather Bureau has developed relationships regarding areal analysis which are presented below along with an explanatory discussion [10]. Any value derived previously in this text is a point value for that particular duration

> which will be equalled or exceeded, on the average, once during the period of interest. For hydrologic design purposes, engineers are more concerned with the average depth of precipitation over an area than with the depth at a particular point. Depth-area curves were developed to meet this need. The depth-area curve is an attempt to relate the average of all point values for a given frequency and duration with a basin to the average depth over the basin for the same duration and frequency. Generally, there are two types of depth-area relations. The first is the storm-centered relation; that is, the maximum precipitation occurring when the storm is centered on the area affected. The second type is the geographically fixed-area relation where the area is fixed and the storm is either centered over it or is displaced so only a portion of the storm affects the area. One can say that storm-centered rainfall data represent profiles of discrete storms, whereas the fixed-area data are statistical averages in which the maximum point values frequently come from different storms. At times, the maximum areal value for the network is from a storm that does not produce maximum point amounts. Each type of depth-area relation is useful, but each must be applied to appropriate data. Generally, the storm-centered relations are used for preparing estimates of probable maximum precipitation, while the geographically fixed relations are used for studies of precipitation-frequency values for basins.

The average depth-area curves in Fig. 4.23 are for fixed areas and were developed by the U.S. Weather Bureau from dense networks. The curves were first prepared for an earlier study during the period 1957-1960 and have since been rechecked against longer-record data [9]. Application of these curves must be consistent with the manner in which they were developed [10]. It is the opinion of both Dr. Robert Clark of the U.S. Weather Service [20] and of the Australian Bureau of Meteorology that these curves are the best currently available.

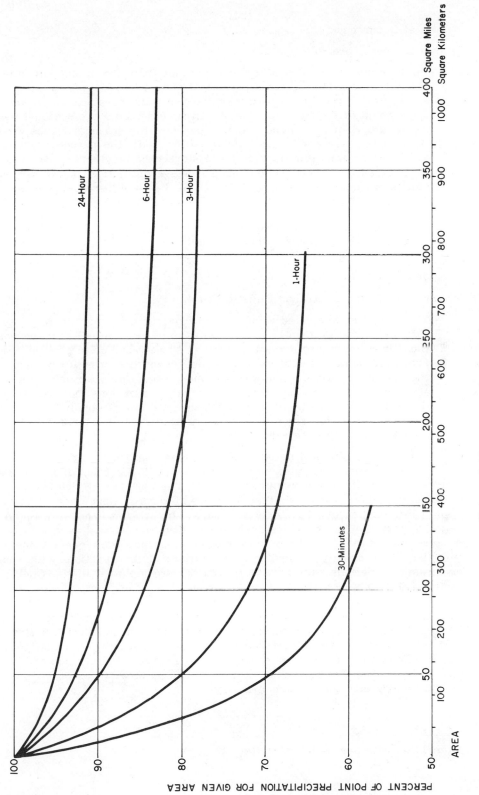

Figure 4.23 Areal analysis graph.

Season of Occurrence

In certain areas, precipitation is highly seasonal.

> Thus, rainy season precipitation-frequency values approach the annual
> values. In sections where the greatest annual N-hour precipitation
> amount may be observed in any season, seasonal precipitation-frequency
> maps would differ from the annual. In no case could the seasonal value
> be greater than the annual value. However, the seasonal values would
> be a certain percent of the annual values, with the percent varying
> according to the frequency of large storms during the season under
> investigation. Generalizations about the season distribution of large
> storms can be obtained from ESSA, *U. S. Weather Bureau Technical
> Paper No. 57* [21].

Currently, there is no convenient manner of applying this knowledge
to the maps of NOAA Atlas 2, other than subjectively.

Temporal Patterns of Storm Rainfall

As indicated in the Australian Bulletin No. 49 [19]:

> The use of sophisticated rainfall-runoff procedure generally requires
> that a temporal pattern be assigned to the design rainfall. In the past,
> these have tended to be assumed either symmetrical or to peak at some
> preferred point of the storm's duration. The intensities for a given re-
> turn period have then been proportioned about this peak using the avail-
> able Intensity-Frequency-Duration curve for durations increasing up to
> the time of concentration. These synthesized curves are not realistic as
> the intensitives for given durations are unlikely to be derived from the
> same period of intense rain.*

The average temporal patterns for the southeastern coastal zone of
Australia for given durations as presented in the 1977 version of Australian
rainfall and runoff are shown in Fig. 4.24. The patterns, which are based
on selected stations with more than 15 years of record, were derived using an
approach described by Pilgrim, et al. [30]. The samples used were taken
from the partial duration series derived from the total record such that the
samples did not overlap. Only those storms which had their peaks coincident
with the most commonly occurring peak position, or for durations with a larger number of periods, adjacent to this peak, are included in the final
sample [31]. These storm temporal patterns are appropriate for use in
the United States where no better information exists, as many similarities have been noted between Australian and North American rainfall characteristics. With increasing catchment area and thus length of critical storms,
it is anticipated that these patterns will be less peaked. When investigating
larger catchments closer attention should be given to the allocation of storm
temporal pattern. For important structures the pattern of past major storms
should be examined for the specific basin.

---

*Reprinted with permission from Ref. 19.

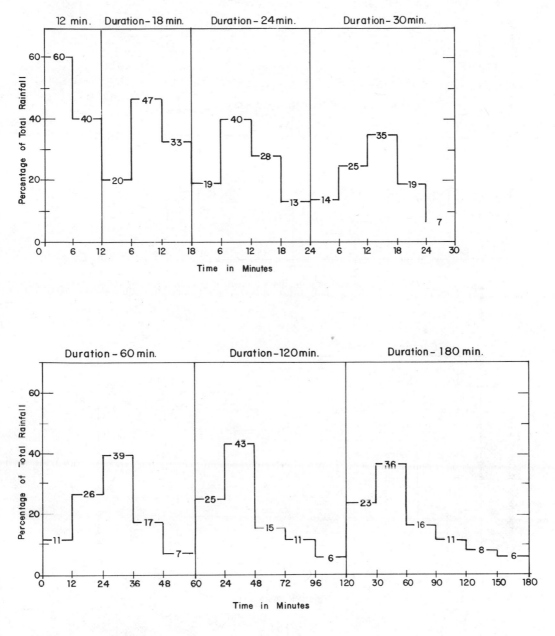

(a)

Figure 4.24 Example temporal pattern of storm rainfall (Melbourne, Australia). (Reproduced with permission from Ref. 19.)

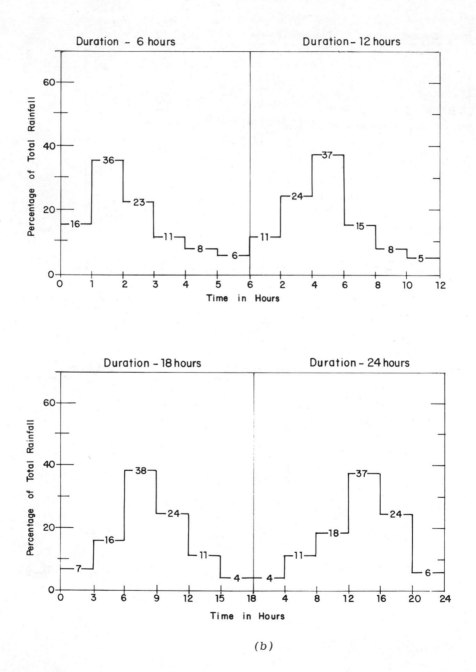

(b)

Figure 4.24 (continued)

4.6 RAINFALL DETERMINATION

This section outlines a procedure for determining the design rainfall event or events to be used for runoff analysis in the drainage planning process. The procedure is also valid for detailed design hydrology; however, simplified methods (that obviate the need for the following procedures) may be used in small basins when Rational Methods are used as explained in Chap. 5 on runoff.

There are several basic points to be considered when determining which rainfall event would cause critical runoff for a particular catchment and with given conditions.

Rainfall Frequencies

Chapter 3 on the drainage planning process provides general guidelines regarding frequencies that should be considered. Drainage planning should always consider the 100-year events as a principal measure for plan effectiveness. Also, several other magnitude events, both smaller and larger, may be considered according to circumstances.

Rainfall Durations

When rainfall durations are approximately equal to the basin response times, generally the runoff for that depth and intensity will be maximized. Several different-duration events should be compared by inputting to the runoff derivation. This will help to identify the duration which results in the highest runoff. Additional durations will need to be considered to determine critical conditions for subbasins if runoff estimates for them are required.

Rainfall Adjustments

Factors have been provided to help the engineer in adjusting rainfall for special effects.

Rainfall Temporal Patterns

Graphs are given in Fig. 4.24 which portray areawide typical rainfall patterns for given duration events.

Note, however, that it is common practice to prepare a design storm by compiling all the various duration characteristics for a given frequency. The incremental amounts are rearranged such that the time of peak rainfall results in maximum runoff.

Rainfall Analysis Steps

The following procedure will help establish rainfall events for a given probability of occurrence that result in maximum runoff. It may be necessary to use several different events with different patterns and characteristics as input to the runoff analysis. This runoff analysis will then help identify the more critical events for the basin under the conditions considered. In cases involving expensive or very important works, it is particularly important to refer to available local data and consult with personnel for additional hydrologic information.

1. Determine the rainfall frequencies of interest according to planning guidelines in Chap. 3.
2. Determine the rainfall durations of interest according to the expected response times of the basin and subbasins involved. Shorter durations than those used for total basins will generally result in more critical situations for the subbasins.
3. For steps 1 and 2 find the appropriate rainfall intensity using the appropriate reference.
4. Using Fig. 4.23 adjust the point rainfall for the areal effect of the basin or subbasin of interest.
5. Summarize and evaluate the significance of step 4. Factors involving changes not greater than 5 percent could be considered insignificant and therefore may be disregarded.
6. For each rainfall duration distribute the rainfall given by the appropriate temporal distribution given in Fig. 4.24 or as described in the section on rainfall temporal patterns. Test other patterns as desired for the local case as available or dictated by the importance and sensitivity of the project.
7. Critical runoff events could be caused by winter, summer, or annual rainfall. There are many possible factors which could cause winter or summer events to be more critical. The determination hinges on the amount of precipitation that actually becomes runoff. This portion of the rainfall is termed *rainfall excess*.* The water that does not run off immediately is held on foliage, stored in surface depressions, or absorbed by the soil. The first two can be categorized as detention losses and the last as infiltration. These losses vary during the year because of vegetation changes, changes in the soil's capacity to infiltrate, and other conditions. Also, some locations may be subject to rapid snowmelt runoff. In such cases, the reader is referred to other publications such as Ref. 22.

4.7 RAINFALL MODELING: THE FUTURE DIRECTION

Due to the typically short nature of hydrologic records, it is usually impossible to accurately assess the adequacy of a hydraulic or water resource facility under extreme or long-term conditions with only the historical records. Hydrologists have for many years used synthetic events in addition to actual records to evaluate designs under more general conditions. For example, in the case of streamflows, long-term synthetically generated records have been used to evaluate the characteristics of storage reservoirs, water supply systems, and irrigation systems.

Where rainfall has been of interest, hydrologists have focused on the creation of synthetic design events. These events were created from the intensity-frequency-duration curves for the location of interest and a desired recurrence interval. But questions such as Where should the peak intensity be located within the storm? Should the storm be of 'triangular' shape? and What is the 50-year storm duration? have never been completely and objectively answered. The construction of a design storm is generally

---

*Rainfall excess is also commonly referred to as effective rainfall or effective precipitation.

recognized as an art and not a science. More important, the use of an N-year return-period rainstorm event may not necessarily produce an N-year return-period runoff event.

Rainfall analysis and generation models are being created and refined to overcome some of the deficiencies of the design storm procedure. The objectives of these models are:

1. To analyze the characteristics of the historical rainfall record at the location of interest
2. To obtain the required statistical parameters that describe rainfall events at this location for synthetic generation purposes
3. To generate long-term synthetic rainfall records that simulate the historical record in its statistical parameters
4. To use this synthetic rainfall record as the input to a simplified run-off model to generate the synthetic runoff record such that the more reliable runoff statistics are known
5. To also create a variety of synthetic storms within this long-term record that could be used to evaluate, through runoff simulation, any given hydraulic/water resource facility

In general, the philosophy behind this approach is that the historical record is only one sample function from the infinite range of possible sequences that could have occurred during that period. Through rainfall modeling, many statistically similar records can be synthesized, each equally likely to occur in the future. This enables the hydrologist to explore a much wider range of possible extreme events to estimate the expected values for these rare events. This procedure allows for much more reliable estimation of extreme hydrologic events such as flood discharge peaks, volumes, or stages [23].

REFERENCES

1. *Dynamic Hydrology*, Peter S. Eagleson, McGraw-Hill, New York, 1970.
2. *Rainfall Intensity-Frequency Data*, David L. Yarnell, Miscellaneous Publication No. 204, U.S. Department of Agriculture, Washington, D.C., 1935.
3. *Rainfall Intensities for Local Drainage Design in the United States for Durations of 5 to 240 Minutes and 2-, 5-, and 10-Year Return Periods*, Weather Bureau Technical Paper No. 24, *Part 1, West of the 115th Meridian*, U.S. Weather Bureau, Washington, D.C., August 1953; revised February 1955; *Part 11, Between 105°W. and 115°W.*, Washington, D.C., August 1954.
4. *Rainfall Intensities for Local Drainage Design in Coastal Regions of North Africa, Longitude 11°W. to 14°E. for Durations of 5 to 240 Minutes and 2-, 5-, and 10-Year Return Periods*, U.S. Weather Bureau, Washington, D.C., September 1954.
5. *Rainfall Intensities for Local Drainage Design in Arctic and Subarctic Regions of Alaska, Canada, Greenland, and Iceland for Durations of 5 to 240 Minutes and 2-, 5-, 10-, 20- and 50-Year Return Period*, U.S. Weather Bureau, Washington, D.C., September 1955.
6. *Rainfall Intensity-Duration-Frequency Curves for Selected Stations in the United States, Alaska, Hawaiian Islands, and Puerto Rico*, U.S. Weather Bureau, Weather Bureau Technical Paper No. 25, Washington, D.C., December 1955.

7. *Rainfall Intensities for Local Drainage Design in Western United States for Durations of 20 Minutes to 24 Hours and 1- to 100-Year Return Period*, U.S. Weather Bureau, Weather Bureau Technical Paper No. 28, Washington, D.C., November 1965.

8. *Rainfall Intensity-Frequency Regime*, U.S. Weather Bureau, Weather Bureau Technical Paper No. 29, Washington, D.C., *Part 1, The Ohio Valley*, June 1957; *Part 2; Southeastern United States*, March 1958; *Part 3, The Middle Atlantic Region*, July 1958; *Part 4, Northeastern United States*, May 1959; *Part 5, Great Lakes Region*, February 1960.

9. *Rainfall Frequency Atlas of the United States for Durations from 30 Minutes to 24 Hours and Return Periods from 1 to 100 Years*, David M. Hershfield, Weather Bureau Technical Paper No. 40, U.S. Weather Bureau, Washington, D.C., May 1961.

10. *Precipitation-Frequency Atlas of the Western United States*, Atlas 2, Volumes I-XI, U.S. Department of Commerce, National Weather Service, National Oceanic and Atmospheric Administration, Silver Spring, Md., 1973.

11. *Climate and Crop Service*, U.S. Weather Bureau, 1896-1910; *Monthly Weather Review*, U.S. Weather Bureau, 1910-1913; *Climatological Data by Sections*, U.S. Weather Bureau, 1914-1964; *Climatological Data by Sections*, Environmental Science Services Administration, 1959-1969; National Climatic Center, Asheville, N. Car.

12. *Hydrologic Bulletin*, U.S. Weather Bureau, 1940-1948; *Hourly Precipitation Data*, U.S. Weather Bureau, 1951-1964; *Hourly Precipitation Data*, Environmental Science Services Administration, 1965-1969; National Climatic Center, Asheville, N. Car.

13. Unpublished data tabulations, California Department of Water Resources, 1900-1969.

14. "Annual Floods and the Partial-Duration Series," W. B. Langbein, *Transactions of the American Geophysical Union*, Vol. 30, No. 6, December 1949, pp. 879-881.

15. "Discussion on Annual Floods and the Partial-Duration Flood Series," by W. B. Langbein and V. T. Chow, *Transactions of the American Geophysical Union*, Vol. 31, No. 6, December 1950, pp. 939-941.

16. *Statistics of Extremes*, E. J. Gumbel, Columbia University Press, New York, 1958.

17. "An Empirical Appraisal of the Gumbel Extreme-Value Procedure," David M. Hershfield and M. A. Kohler, *Journal of Geophysical Research*, Vol. 65, No. 6, June 1960, pp. 1737-1746.

18. "An Empirical Comparison of the Predictive Value of Three Extreme-Value Procedures," David M. Hershfield, *Journal of Geophysical Research*, Vol. 67, No. 4, April 1962, pp. 1535-1542.

19. *Point Rainfall Intensity-Frequency-Duration Data, Capital Cities*, C. L. Pierrehumbert, Department of Science, Bureau of Meteorology, Bulletin No. 49, Australian Government Publishing Services, Canberra, August 1974.

20. Private communication, Robert Clark, U.S. Weather Services, National Oceanic and Atmospheric Administration, 1976.

21. *Normal Monthly Number of Days with Precipitation of 0.5, 1.0, 2.0, and 4.0 Inches or More in the Conterminous United States*, J. F. Miller and R. H. Frederick, Weather Bureau Technical Paper No. 57, U.S. Environmental Science Services Administration, U.S. Weather Bureau, Washington, D.C. 1966.

22. *Snow-Hydrology*, U.S. Army Corps of Engineers, Portland Ore., June 30, 1956.

23. *MITCAT Catchment Simulation Model, Description and Users Manual*, Resource Analysis, Inc., Version 6, Cambridge, Mass., May 1975.

24. *Urban Storm Drainage Criteria Manual*, Volume I, Wright-McLaughlin Engineers, Denver Regional Council of Governments, Denver, 1969; revised May 1975.

25. *Five to 60 Minute Precipitation Frequency Data for Eastern and Central United States*, National Oceanic and Atmospheric Administration Technical Memorandum NWS Hydro-35, Silver Spring, Md., 1977.

26. *Probable Maximum Precipitation Estimates, United States East of the 105th Meridian*, National Oceanic and Atmospheric Administration Hydrometeorological Report No. 51, Washington, D.C., June 1978.

27. *Probable Maximum Precipitation Estimates, Colorado River and Great Basin Drainages*, National Oceanic and Atmospheric Administration Hydrometeorological Report No. 49, Silver Springs, Md., September 1977.

28. *Design of Small Dams*, 2nd ed., U.S. Bureau of Reclamation, Washington, D.C, 1973.

29. *Short Period Rainfall Intensity Analysis*, C. L. Pierrehumbert, Working Paper 156, Bureau of Meteorology, Australia, 1972.

30. "Temporal Patterns of Design Rainfall for Sydney," D. H. Pilgrim, I. Cordery, and R. French, *Civ. Eng. Trans. I.E., Australia, CE11, 1,* 1969, pp. 9-14.

31. "Design Temporal Patterns of Storm Rainfall in Australia," A. J. Hall and T. H. Knoon, *Hydrol. Symp.,* Perth, I.E. Australia, 1973, pp. 77-84.

5

Runoff Hydrology: Methodology

5.1 INTRODUCTION

The storm runoff peak, volume, and timing provide the basis for all floodplain definition and planning, design, and construction of drainage facilities. To be in error on the hydrology means that floodplain maps are unreliable or that works are either undersized, oversized, or out of hydraulic balance. In floodplain management and urban drainage design it is volume which holds the key to management and cost optimization. The intent of this chapter is to summarize dependable methods of approximating the characteristics of rainfall runoff. Design rainfall was discussed in Chap. 4. Examples of techniques are presented in Chap. 6.

Analytical Methods

The review of current practice coupled with the need for better approximations of storm runoff magnitude has shown that four basic approaches can be utilized for determining the character of storm runoff. They are the Rational Method, the Synthetic Unit Hydrograph Procedure (SUHP), computer simulation modeling, and statistical analyses.

This chapter presents general descriptions of the methods used to analyze and predict runoff peak flows or hydrographs. The following chapter presents several example techniques that are used on a daily basis in the engineering profession to quantify these flows or hydrographs.

Applicability of Methods

For most drainage basins, the many basin and subbasin characteristics and existing or probable development patterns will necessitate the use of the SUHP or computer simulation modeling techniques. These deal directly with volume of flow as well as peak rate of flow, both of which are important in comprehensive drainage engineering work. Drainage basin management is a space allocation problem which emphasizes the need to analyze for the time distribution of storm water runoff and its volume.

For drainage basins that are uncomplicated the Rational Method is suggested for planning and design purposes. The Rational Method is in use throughout the world and has been found to be satisfactory for smaller and simpler drainage systems.

5.2 RATIONAL METHOD

The Rational Method is used in most engineering offices. It has frequently come under criticism for its simplicity, but no other practical drainage design method has evolved to a greater level of general acceptance by the practicing engineer. The Rational Method, properly understood and applied, produces satisfactory results for urban storm drainage design within certain limits [1-4].

Rational Formula

The Rational Method is based on the rational formula:

$$Q = 0.275 \text{ CIA metric} \qquad [Q = CIA \text{ British}] \qquad (5.1)$$

Q is defined as the maximum rate of runoff in cubic meters per second [cubic feet per second]. Actually, Q in British units is in terms of inches per hour per acre; however, since this rate differs from cubic feet per second by less than 1 percent, the latter is used. C is a runoff coefficient which is the ratio between the maximum rate of runoff from the area A in terms of square kilometers [acres] and the average rate of rainfall intensity I in millimeters per hour [inches per hour] for the period of maximum rainfall of a given frequency of occurrence having a duration equal to the time of concentration. The time of concentration usually is the time required for water to flow from the most remote point of the area to the point being investigated.

Assumptions

The basic assumptions made when the Rational Method is applied are:

The computed maximum rate of runoff to the design point is a function of the average rainfall rate during the time of concentration.
The maximum rate of rainfall occurs during the time of concentration, and the design rainfall depth during the time of concentration is converted to the average rainfall intensity for the time of concentration.
The maximum runoff rate occurs when the entire area is contributing flow, which is defined as the time of concentration.

Limitations

The Rational Method is adequate for approximating the peak rate of runoff from a rainstorm in a given basin within certain limits. The greatest drawback to the Rational Method is that it normally provides only this one point on the runoff hydrograph. The Rational Method was not developed for the study of runoff volume, although some approximations have been developed. When the basins become complex and where subbasins come together, the Rational Method will tend to overestimate the actual flow, which may result in over-sizing drainage facilities. The Rational Method should normally be used only for basins of 40 hectares [100 acres] or less.

5.3 RAINFALL EXCESS AND INFILTRATION

*Rainfall excess** is that portion of the precipitation which appears in surface streams and man-made subsurface channels during and after a rainstorm. Those portions of precipitation that do not reach the channels are called abstractions, and include: interception by vegetation, evaporation, infiltration, storage in all surface depressions, and long-time surface detention. The total design rain falling to the earth can be obtained from Chap. 4 on rainfall. This section illustrates methods for determining the amount of rainfall that actually becomes runoff.

The methods described in this section are for use with the Synthetic Unit Hydrograph Procedure (SUHP) and computer modeling and do not apply to the Rational Method because in the Rational Method the abstractions are accounted for in the C factor. However, the information presented here can aid the engineer in selecting a reasonable runoff coefficient for the Rational Method.

There are many representations of rainfall losses which can be used to arrive at rainfall excess, though none are completely satisfactory [5-11]. Only a few of these methods are presented herein.

Pervious-Impervious Areas

All parts of a basin can be considered either pervious or impervious. The *pervious* part of a drainage basin is that area where water can readily infiltrate into the ground. The *impervious* part is the area that does not readily allow water to infiltrate into the ground, such as areas that are paved or covered with buildings and footpaths. In urban hydrology, the percent of pervious and the percent of impervious land are important. As urbanization occurs, the percent of impervious area increases and the rainfall runoff changes significantly, with the total amount of runoff normally increasing, the time of runoff decreasing, and the peak runoff rates increasing substantially [12, 13].

When analyzing an area for design purposes, the probably future percent of impervious area must be estimated. Table 5.1 is presented as a guide.

---

*Rainfall excess is also commonly referred to as effective rainfall or effective precipitation.

Table 5.1 Land Use vs. Percent of Perviousness/Imperviousness

Land use	Percent pervious	Percent impervious
Central business zone area, shopping centers, etc.	0- 5	95-100
Residential:		
Dense (apartment houses)	40- 55	45- 60
Normal (detached houses)	55- 65	35- 45
Large lots	60- 80	20- 40
Parks, greenbelts, passive recreation	90-100	0- 10

Table 5.2 Typical Depression and Detention for Various Land Covers

Land cover	Range (mm)	Range (in.)	Recommended (mm)	Recommended (in.)
Impervious:				
Large paved areas	1.3- 3.8	0.05-0.15	2.5	0.1
Roofs, flat	2.5- 7.5	0.1 -0.3	2.5	0.1
Roofs, sloped	1.2- 2.5	0.05-0.1	1.2	0.05
Pervious:				
Lawn grass	2.0-12.5	0.1 -0.5	7.5	0.3
Wooded areas and open fields	5.0-15.0	0.2 -0.6	Assess each situation	

Depression and Detention Losses

Rainwater that is collected and held in small depressions and does not become part of the general runoff is called *depression storage*. Most of this water eventually infiltrates or evaporates. *Detention losses* include water intercepted by trees and bushes and water that is retained and detained on the surface. The water that is held in depressions on roofs and roads and eventually evaporates is considered a part of depression and detention losses. Table 5.2 can be used as a guide for estimating the amount of depression and detention storage.

When an area is analyzed for depression and detention storage, the various pervious and impervious storage values must be considered according to the percent of area coverage.

There are other losses in the impervious area that are not readily quantified such as water lost to evaporation off warm surfaces and water lost due to the natural amount of water that attaches to the surface and cannot run off. This is referred to here as a general pervious area loss. An amount of 2.5 mm (0.1 in.) will typically be used.

Infiltration

The penetration of water through the soil surface is called *infiltration*. In urban hydrology much of the infiltration occurs on areas covered with lawns and gardens. Urbanization normally decreases the total amount of infiltration [14].

Soil type is an important factor in determining the infiltration rate. When the soil has a large percent of well-graded fines, the infiltration rate is low. In some cases of extremely tight soil there may be, from a practical standpoint, essentially no infiltration. If the soil has several layers or horizons, the least permeable layer will sometimes control the steady infiltration rate. The infiltration rate is the rate at which water enters the soil at the surface and which is controlled by surface conditions, and the transmission rate is the rate at which the water moves in the soil and which is controlled by the horizons.

During a rainstorm, the infiltration rate generally decreases with time. The hydrologist has several methods available to consider and simulate this phenomena.

The soil cover also plays an important role in determining the infiltration rate. Vegetation, lawn grass in particular, tends to increase infiltration by loosening the soil near the surface and ponding water so that the hydraulic head created helps force the water into the ground.

Antecedent precipitation can satisfy wholly or partially the higher initial infiltration. When rainfall occurs on an area that has little antecedent moisture, that is, the ground is dry, the infiltration rate is much higher than it is with a high antecedent moisture such as from a previous storm or from irrigation.

A high-intensity storm can, in some cases, affect the soil surface sufficiently to cause a change in the infiltration.

Other factors affecting the infiltration rates include: slope of land, temperature, quality of water, age of lawn, and soil compaction [8].

Field testing may be useful in determining relative infiltration characteristics for basins being analyzed for storm runoff. The number of tests would depend on the extent of the basin and the homogeneity of the soil and cover type.

There are several acceptable methods for determining relative infiltration rates in the field [15]. The simplest consists of an 18-in.-diameter infiltration ring which is driven into the soil. Water is poured into the ring until it is almost full. Then the drop in water surface is measured at various time intervals. Water can then be added in increments or automatically kept at a constant level by a simple piping and filling tank arrangement. For dependability, the tests should proceed until the infiltration rate is essentially constant.

Although it is desirable to have each basin being analyzed for storm runoff field-tested for its specific pattern of infiltration, guideline infiltration values for preliminary storm runoff analysis and sewer design can be made. This value is 12.5 mm/hr [0.5 in./hr], expressed as a constant value.

Infiltration may be represented mathematically by one of several different functions. The variables in these functions can be assigned by analysis of infiltration tests and various data regarding soils, geology, vegetation, and other parameters mentioned previously. Some of these functions are summarized and explained in detail in the following section.

Representation of Infiltration

Although infiltration research has been conducted for many years, there still is no generally recognized adequate quantitative model of natural infiltration. Considering the complex combination of soil characteristics, soil moisture conditions, and other factors occurring in nature, that is not too surprising. Nevertheless, infiltration is extremely important because it can influence not only the volume and intensity of rainfall excess rates but the timing of the runoff hydrograph as well. Some alternate representations of infiltration are as follows:

Horton's equation
Holtan's Method
Soil Conservation Service (SCS) Method
Algorithms based on Darcy's law and/or antecedent conditions
Simple Abstraction Method

Each of these methods permits initial infiltration rates to depend on some measure of antecedent soil moisture and to vary with time during the storm.

5.4 SYNTHETIC UNIT HYDROGRAPH PROCEDURE (SUHP)

For basins that are larger than about 40 hectares [100 acres], for some complex basins that are smaller than this, and for problems where determination of volume or timing is critical, it is recommended that the design storm runoff be analyzed by deriving synthetic unit hydrographs or methods other than the Rational Method. The hydrograph procedure provides a volume distribution time and therefore affords the opportunity to better manage the runoff in a cost-effective manner. The unit hydrograph principle was originally developed by Sherman in 1932 [16]. The synthetic unit hydrograph, which is used for analysis when there is no rainfall runoff data for the basin under study, was developed by Snyder in 1938 [17]. Since that time, there have been many developments and modifications to these methods which basically derive varying empirical relations but have not significantly modified the basic concepts [7, 8, 10, 18-27].

Definition

A *unit hydrograph* is defined as the response of 1 unit (usually an inch) of direct runoff from a drainage area. A unit storm is a rainfall of such duration that the period of surface runoff is not appreciably less for any rain of shorter duration. This unit hydrograph represents the integrated effects of factors such as tributary area, shape, street pattern, channel capacities, and street and land slopes [10, 20-22].

The runoff response (design hydrograph) to a design storm is determined by multiplying the ordinates of the unit hydrograph by incremental rainfall excess depths of the design storm and summing each of the resulting incremental runoff hydrographs sequentially with respect to time. This is illustrated in Fig. 5.1.

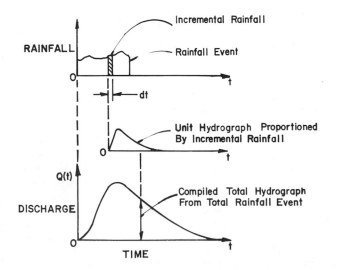

Figure 5.1 Synthetic hydrograph demonstration.

A unit hydrograph for a watershed may be derived by analyzing actual rainfall and runoff data. Often, this type of analysis cannot be completed because of a lack of data, and it has limited value because it allows no direct prediction of the effects of urbanization. However, empirical relationships have been derived by analysis of hydrographs from many different kinds of watersheds. These relationships lead to synthetic unit hydrographs for basins without historical data and for various development conditions.

Basic Assumptions

The derivation and application of the unit hydrograph are based on the following assumptions:

> The rainfall intensity is constant during the storm that produces the unit hydrograph.
> The rainfall is uniformly distributed throughout the entire area of the drainage basin.
> The base or time duration of the design runoff due to rainfall excess of unit duration is constant.
> The ordinates of the design runoff with a common base time are directly proportional to the total amount of direct runoff represented by each hydrograph.
> The effects of all physical characteristics of a given drainage basin, including shape, slope, detention, infiltration, drainage pattern, channel storage, etc., are reflected in the shape of the unit hydrograph for that basin.

Routing

Streamflow routing of runoff can substantially modify discharge hydrographs.

A good procedure to use when performing hydrologic calculations is to subdivide a drainage basin of interest into a number of small subbasins of 2.5 to 8 km^2 [1 to 3 mi^2] that have similar character. Hydrographs should be calculated for each subbasin and the resulting hydrographs should then be routed down the main stream channels. There are numerous routing methods which are best referred to directly [7, 8, 10, 28].

5.5 COMPUTER MODELING APPROACHES

During the last decade a number of hydrologic and/or water quality computer models have been developed and are in practical use in engineering. The advanced computer models allow the use of sophisticated algorithms to represent various phenomena and complicated flow routing. Without computer models, such algorithms could not be practically used. These computer models have made possible a better understanding and analysis of runoff from both urban development and rural land. The end result of computer modeling in the case of drainage is an improved runoff system.

Various publications are available that describe the methods and results of tests of these methods [29-32].

The publication *Evaluation of Mathematical Models for the Simulation of Time-Varying Runoff and Water Quality in Storm and Combined Sewerage Systems* [32] presented the results of a comprehensive review of most computer models available in North America and Europe. Table 5.3 presents in

Table 5.3 Comparison of Major Model Categories.

Model	\<CATCHMENT HYDROLOGY\> Multiple catchment inflows	Dry-weather flow	Input of several hyetographs	Snowmelt	Runoff from impervious areas	Runoff from pervious areas	Water balance between storms	\<SEWER HYDRAULICS\> Flow routing in sewers	Upstr and downstr flow control	Surcharging and pressure flow	Diversions	Pumping stations	Storage	Prints stage	Prints velocities	\<WASTEWATER QUALITY\> Dry-weather quality	Stormwater quality	Quality routing	Sedimentation and scour	Quality reactions	Wastewater treatment	Quality balance between storms	Receiving water flow simulation	Receiving water quality simulation	\<MISCELLANEOUS\> Continuous simulation	Can choose time interval	Design computations	Real-time control	Applied to real problems	Computer program available
BATTELLE NORTHWEST	●	●	●		●	●		●		●	●		●	●	●	●	●	●			●						●	●	●	●
BRITISH ROAD RESEARCHLAB	●	●	●	●	●	●	●	●															●			●			●	●
CHICAGO FLOW SIMULATION	●	●		●	●	●	●	●					●	●	●								●		●	●			●	●
CHICAGO HYDRO-GRAPH METHOD	●	●		●	●	●	●	●	●		●				●								●			●	●		●	●
COLORADO STATE UNIVERSITY	●	●			●	●		●							●											●				●
CORPS OF ENGINEERS	●	●	●		●	●	●	●	●		●		●		●											●	●		●	
DORSCH CONSULT	●	●			●	●	●	●	●	●	●	●	●	●	●		●	●		●		●				●			●	●
ENV PROTECTION AGENCY	●	●	●	●	●	●	●	●	●	●	●	●	●	●	●	●	●	●	●	●	●	●	●	●	●	●	●	●	●	●
HYDROCOMP	●	●			●	●		●	●	●	●	●	●	●	●		●	●					●		●	●			●	
MASSACHUSSETTS INST OF TECHN	●	●			●	●	●	●	●	●	●	●	●	●			●	●					●			●			●	
MINNEAPOLIS-ST. PAUL	●	●			●	●		●	●	●	●	●	●	●	●											●		●	●	●
SEATTLE	●				●	●		●	●	●	●	●	●	●	●			●		●			●	●		●		●	●	●
SOGREAH	●	●			●	●	●	●	●	●	●	●	●	●	●											●			●	
UNIVERSITY OF CINCINNATI	●	●	●		●	●		●	●					●	●											●				
UNIVERSITY OF ILLINOIS	●			●	●	●		●					●	●												●	●		●	●
UNIVERSITY OF MASSACHUSSETS		●			●	●	●	●						●	●								●	●	●	●			●	
WATER RESOURCES ENGINEERS	●	●	●		●	●		●	●				●	●	●	●	●	●		●			●	●		●	●		●	
WILSEY AND HAM	●	●			●	●		●																			●		●	●

Source: Reproduced from Ref. 32.

matrix form a summary of the capabilities of 18 computer models. The authors of this paper also arrived at the following recommendations regarding routine applications:

> Various models stand out due to their completeness of hydrologic and hydraulic formulations, the ease of input data preparation, the efficiency of computational algorithms, and the adequacy of the program output. Other models, although deficient in some of these respects, merit consideration due to special features which are not included in the more comprehensive models but may be required for specific applications.

The following models are consequently recommended for routine applications:

1. Battelle Urban Wastewater Management Model for real-time control and/or design optimization considering hydraulic, water quality and cost constraints, provided the hydrologic and hydraulic model assumptions are adequate for particular applications (lumping of many small subcatchments into few large catchments, neglect of downstream flow control, backwater, flow reversal, surcharging, and pressure flow).

2. Corps of Engineers STORM Model for preliminary planning of required storage and treatment capacity for storm runoff from single major catchments, considering both the quantity and quality of the surface runoff and untreated overflows.

3. Dorsch Consult Hydrograph Volume Method for single-event flow analysis considering most important hydraulic phenomena (except flow reversal). A Quantity-Quality Simulation Program for continuous wastewater flow and quality analysis is now available, but the model was completed too late for evaluation.

4. Environmental Protection Agency Stormwater Management Model for single-event wastewater flow and quality analysis provided the hydraulic limitations of the model are acceptable (neglect of downstream flow control and flow reversal, inadequate backwater, surcharging, and pressure flow formulation). A new version patterned after the Corps of Engineers STORM Model is now available for continuous simulation, but this version was completed too late for evaluation.

5. Hydrocomp Simulation Program for single-event and continuous wastewater flow and quality analysis provided the hydraulic limitations of the model are acceptable (approximate backwater and downstream flow control formulation, neglect of flow reversal, surcharging, and pressure flow).

6. Massachusetts Institute of Technology Urban Watershed Model for single-event flow analysis provided the hydraulic limitations of the model (neglect of backwater, downstream flow control, backwater, flow reversal, surcharging, and pressure flow). The use of a separate model for these phenomena is acceptable.

7. Seattle Computer Augmented Treatment and Disposal System as an example of an operating real-time control system to reduce untreated overflows. A more comprehensive computer model simulating both wastewater flow and quality and including mathematical optimization should be considered, however, for new systems.

8. SOGREAH Looped Sewer Model for single-event wastewater flow and quality analysis considering all important hydraulic phenomena.

9. Water Resources Engineers Stormwater Management Model for single-event wastewater flow and quality analysis considering all important hydraulic phenomena.

The hydrologist and other technical staff members should carefully evaluate which model would be appropriate for a particular engineering investigation. No one computer model can reasonably meet the needs of all hydrological investigation types. A mix of a few of the models would be required to meet such a comprehensive demand.

The following list tabulates references for some additional computer models. Refer to available references for complete descriptions and tests of these models.

CURM   Cincinnati Urban Runoff Model [33]
HEC-1   Hydrologic Engineering Center, U.S. Army Corps of Engineers [28]
RRL   Road Research Laboratory Hydrograph Method [34, 35]
SCSH   Soil Conservation Service– Project Formulation–Hydrology [36]

These models can give good results and test a number of possible solutions to drainage problems when used in a professional manner.

The following chapter contains a description of the Massachusetts Institute of Technology Catchment Model (MITCAT) to illustrate the workings and input data for a modern deterministic model utilizing a kinematic wave formulation.

MITCAT is one of the models that has been used extensively as part of comprehensive drainage planning programs. Its strengths lie in use of deterministic runoff routing, ability to analyze complex drainage basins and reservoir linkages, ease of data acquisitional input, availability of many different infiltration algorithms, and reasonable user cost.

5.6 STATISTICAL ANALYSIS

For larger drainage areas the experienced hydrologist prefers to compute flood flows for a given return period using actual records of discharges (which have been gaged) whenever available. The results are usually more dependable than those obtained by the use of the Rational Method, Unit Hydrograph Method, or other approaches, if the period of record is long enough.

In urban hydrology, the preferred statistical approach is limited by the almost total lack of adequate runoff records in urban areas, the effect of rapid urbanization, and because study areas which have been gaged for a satisfactory period usually have records which represent undeveloped basins. Once an area undergoes urbanization, the records representing natural conditions no longer apply to future conditions. Thus, use of the Rational Method and the SUHP or computer modeling will generally be necessary in the region for urban or urbanizing areas.

The statistical analysis approach has its greatest applicability to natural streams, the basins of which will remain in a natural state. Such streams include those with large basins where the urbanization effect on runoff will be negligible and small streams where the basin primarily consists of undevelopable land or greenbelt areas.

The logic involved in the statistical approach to determining the size of flood peaks is that nature, over a period of years, has defined a flood magnitude-frequency relationship which can be derived by study of actual occurrences. A period of record on a particular basin where the floods have been measured and recorded is considered to be a representative period, and floods which occurred during the period can be assumed to occur in a similar future period; that is, the period may be expected to repeat itself.

The engineer will generally want to approximate the flood which will occur once in 100 years on a given stream. If a period of record of 25 years is available, there are various sophisticated and complex computation procedures available to extrapolate the 25-year period of record to 100 years [10]. However, the extrapolation can be nothing more than an approximation because of the vagaries of nature.

A statistical procedure which was chosen as a standard in 1968 by the Water Resources Council of the U.S. government is the log-Pearson Type III.

REFERENCES

1. *Introduction to Hydrology*, W. Viessman, T. Harbaugh, and J. Knapp, Intext Educational Publishers, New York, 1972.
2. *Design and Construction of Sanitary and Storm Sewers*, ASCE Manual of Engineering Practice, No. 37, American Society of Civil Engineers, New York, 1958.
3. "Determination of Runoff for Urban Storm Water Drainage System Design," Kurt W. Bauer, *Technical Record*, Vol. 2. Nos. 4 and 5, April-May 1965.
4. "Determination of Run-off Coefficients," *Public Works*, Vol. 94, No. 11, Ridgewood, N.J., November 1963.
5. *Surface Runoff Phenomena, Part I, Analysis of the Hydrograph*, R. E. Horton, Horton Hydrological Laboratory, Publ. 101, Ann Arbor, Mich., 1935.
6. *Model of Watershed Hydrology*, USDA HL-70, Technical Bulletin No. 1435, U.S. Department of Agriculture, Agricultural Research Service, Washington, D.C., 1971.
7. *Hydrology for Engineers*, Ray K. Linsley, Jr., Max A. Kohler, and Joseph L. H. Paulhus, McGraw-Hill, New York, 1958.
8. "Hydrology, Part I, Watershed Planning," *National Engineering Handbook*, Section 4, *Hydrology*, U.S. Department of Agriculture, Soil Conservation Service, Washington, D.C., August 1972.
9. "Modeling Infiltration during a Steady Rain," R. G. Mein and C. L. Larson, *Water Resources Research*, Vol. 9, 1973, pp. 384-394.
10. *Handbook of Applied Hydrology*, Ven Te Chow, McGraw-Hill, New York, 1964.
11. *Dynamic Hydrology*, Peter S. Eagleson, MIT, McGraw-Hill, New York, 1970.
12. *Hydrology for Urban Land Planning—A Guidebook on the Hydrologic Effects of Urban Land Use*, Luna B. Leopold, Geological Survey Circular 554, U.S. Department of the Interior, Geological Survey, Washington, D.C., 1968.
13. *Modern Hydrology*, Raphael G. Kazmann, Louisiana State University, Harper & Row, New York, 1972.
14. *Urban Growth and the Water Regimen*, John Savini and J. C. Kammerer, Geological Survey Water-Supply Paper 1591-A, U.S. Department of the Interior, Geological Survey, Washington, D.C., 1961.

15. *A Field Method for Measurement of Infiltration*, A. U. Johnson, Geological Survey Water-Supply Paper 1544-F, U.S. Department of the Interior, Geological Survey, Washington, D.C., 1963.

16. "Stream Flow from Rainfall by the Unit-Graph Methods," L. K. Sherman, *Engineering News Records*, Vol. 108, New York, April 7, 1932.

17. "Synthetic Unit-Graphs," F. F. Snyder, *Transactions of the American Geophysical Union*, Vol. 19, Washington, D.C., 1938.

18. *Urban Hydrology for Small Watersheds*, Technical Release No. 55, U.S. Department of Agriculture, Soils Conservation Service, Engineering Division, Washington, D.C., January 1975.

19. *Unit Hydrograph*, Civil Works Investigations, Project 152, U.S. Army Engineer District, Baltimore, Corps of Engineers, Baltimore, 1963.

20. *Unit Hydrograph Characteristics for Sewered Areas*, Peter S. Eagleson, Proceedings ASCE, HY2, American Society of Civil Engineers, New York, March 1962.

21. *Unitgraph Procedures*, U.S. Department of the Interior, Bureau of Reclamation, Denver, November 1952; revised August 1965.

22. *Flood-Hydrograph Analyses and Computations Manuals*, U.S. Army Corps of Engineers, EM 1110-2-1405, Washington, D.C., August 31, 1959.

23. *Urban Storm Drainage Criteria Manual*, Volumes I and II, Denver Regional Council of Governments, Wright-McLaughlin Engineers, Denver, 1969; revised May 1975.

24. *The Application of Synthetic Unit Hydrographs to Drainage Basins in the Riverside County Flood Control and Water Conservation District*, Riverside, Calif., January 1963.

25. *Determination of Urban Watershed Response Time*, E. F. Schulz and O. G. Lopez, Hydrology Paper 71, Colorado State University, Fort Collins, December 1974.

26. *Volume and Rate of Runoff in Small Watersheds*, SCS-TP-149, U.S. Department of Agriculture, Soil Conservation Service, Washington, D.C., Janaury 1968.

27. *A Study of Some Effects of Urbanization on Storm Runoff from a Small Watershed*, Report 23, Texas Water Development Board, University of Texas, Austin, August 1966.

28. *HEC-1, Flood Hydrograph Package, Users Manual*, U.S. Army Corps of Engineers, Washington, D.C., 1959.

29. *A Critical Review of Currently Available Hydrologic Models for Analysis of Urban Stormwater Runoff*, Contract No. 14-31-0001-3416, R. K. Linsley, for the Office of Water Resources Research, Palo Alto, Cal., August, 1971.

30. *Models and Methods Applicable to Corps of Engineers Urban Studies*, J. W. Brown, M. R. Walsh, R. W. McCarley, A. J. Green, Jr., and H. W. West, U.S. Army Engineer Waterways Experiment Station, Miscellaneous Paper H-74-8, Vicksburg, Miss., August 1974.

31. *Assessment of Mathematical Models for Storm and Combined Sewer Management*, A. Brandstetter, Environment Protection Technology Series, EPA-600/2-76-175a, Municipal Environmental Research Laboratory, Cincinnati, Ohio, August 1976.

32. *Evaluation of Mathematical Models for the Simulation of Time-Varying Runoff and Water Quality in Storm and Combined Sewerage Systems*, A. Brandstetter, R. Field, and H. Torno, paper presented at the conference on Environmental Modeling and Simulation, sponsored by the U.S. Environmental Protection Agency, Edison, N.J., April 1976.

33. *Urban Runoff Characteristics*, Division of Water Resources, Department of Civil Engineering, University of Cincinnati, Water Pollution Control Research Series, 11024 DQU 10/70, Environmental Protection Agency, Water Quality Office, Washington, D.C., 1970.

34. *The Design of Urban Sewer Systems*, L. H. Watkins, Department of Scientific and Industrial Research, Road Research Laboratory, Technical Paper No. 55, London, 1962.

35. *Developments in Urban Hydrology in Great Britain*, L. H. Watkins and C. P. Young, Laboratory Note No. Ln/885-LHW.CPY, Conference on Urban Hydrology Research, Proctor Academy, Andover, N.H., August, 1965.

36. *Computer Program for Project Formulation-Hydrology*, Technical Release No. 20, U.S. Department of Agriculture, Engineering Division, Soils Conservation Service, Washington, D.C., May 1965.

37. *Assessment of Mathematical Models for Storm and Combined Sewer Management*, EPA Technology Series, EPA 600/2-76-175a, Cincinnati, Ohio, August 1976, p. 4.

6
Runoff Hydrology: Example Techniques

6.1 INTRODUCTION

The basic methodology presented in the previous chapter is implemented by the use of a combination of hydrology techniques. This chapter presents several of these hydrology techniques with primary task descriptions as shown in Table 6.1.

6.2 RATIONAL METHOD

Introduction

The Rational Method of estimating runoff flows is explained in the following paragraphs. This method generally gives adequate results for basins of 40 hectares (100 acres) or less.

The basic algorithm is:

$$Q = 0.275CIA \text{ metric} \quad [CIA \text{ British}] \tag{6.1}$$

where C = runoff coefficient
I = rainfall intensity, mm/hr [in/hr]
A = area, km^2 [acres]

Briefly, the methodology used to select these various parameters is as follows:

The time of concentration t_c for the basin is analyzed and selected.
The average rainfall intensity I for a duration equal to the time of concentration is derived as discussed in Chap. 4 on rainfall for the frequency of interest.
The tributary area A is measured.
The runoff coefficient C is selected based upon guidelines and comparative examination of the tributary basin.
Adjustments for infrequent storms are made where appropriate.
Peak discharge is estimated according to the rational formula.

Time of Concentration

One of the basic assumptions underlying the Rational Method is that runoff is a function of the average rainfall rate during the time required for water to

Table 6.1 Example Hydrology Techniques

Task description	Technique
Peak flow determination	Rational Method
Rainfall excess determination	Horton's equation
Rainfall excess determination	Holtan's Method
Rainfall excess determination	SCS Method
Synthesis of runoff hydrographs	Synthetic unit hydrograph procedures
Computer modeling	MITCAT
Statistical peak flow analysis	Log-Pearson III

flow from the most remote part of the drainage area under consideration to the point of interest. This time is called the time of concentration. The rainfall intensity I is determined from rainfall intensity-frequency curves for a duration equal to the time of concentration.

For urban storm drains, the time of concentration consists of an inlet time (or time required for runoff to flow over the surface to the nearest inlet or point of converged streamflow) and the time of flow in the drain to the point of interest. The latter time can be estimated from the hydraulic properties of the drain. Inlet time will vary with surface slope, depression storage, surface cover, antecedent rainfall, and infiltration capacity of the soil, as well as distance of surface flow. In general, the higher the rainfall intensity, the shorter the inlet time. Common urban practice is to vary the inlet time from 10 to 30 min. It should be noted that the time of concentration has no relationship to the time of beginning of rainfall, being related rather to the position of the maximum rainfall intensity. When dealing with pipe systems, the time of concentration may be readily calculated from the inlet time plus time of flow in each successive pipe run. The latter value is calculated from the velocity of flow as given by Manning's formula for hydraulic conditions prevailing in the pipes.

The inlet time can be estimated by calculating the various overland distances and flow velocities from the most remove point. A common mistake is to assume velocities that are too high for the areas near the storm-collecting drains. Often the remote areas have flow that is very shallow, and in this case the velocities cannot be calculated by channel equations such as Manning's, but special overland flow analyses must be considered [1]. Figure 6.1 can be used to help estimate time of overland flow [2].

Another common error is to analyze only the flow from the entire basin when a smaller portion of the basin has quicker response and a higher proportion of rainfall that becomes runoff, and thus higher peak flow rates. This situation is often encountered in a long basin, or a basin where the upper portion contains rural areas or grassy parkland and the lower portion is developed urban land. Thus flowrates from homogenous subbasins having high runoff potential should also be analyzed.

Figure 6.1 is also a guide to be used for estimating the flow times in street channels and pipes. The drainage network characteristics should be carefully checked to determine if they are within the range of and appropriately represented by this graph.

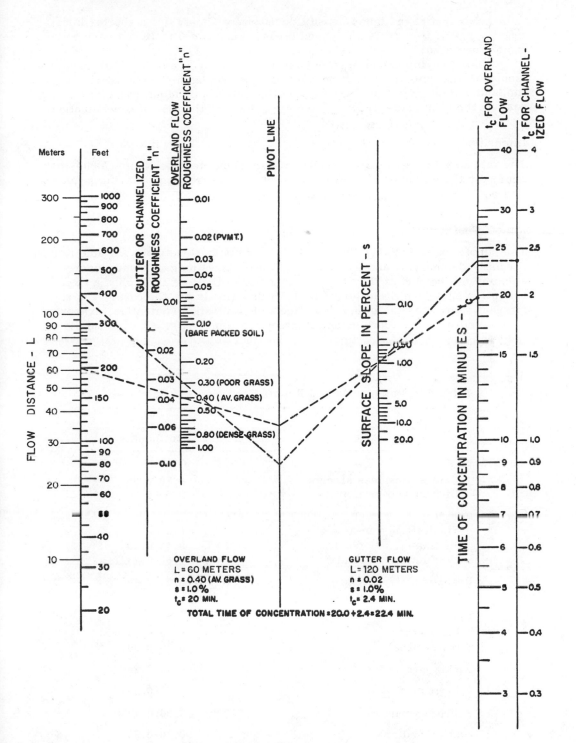

Figure 6.1 Nomograph for time of concentration.

Areas that have higher density development should use a shorter inlet time for runoff from roofs and paved areas. Values of 5 to 10 min have commonly been used.

When studying proposed subdivision land the overland flow path perpendicular to the contours should not be used for calculations. The reason is that the land will be graded and swales will often intercept the natural contour and conduct the water to the streets, thus cutting down on the time of concentration.

Intensity

The intensity I is the average rainfall rate in millimeters per hour (inches per hour) for the period of maximum rainfall of a given frequency having a duration equal to the time of concentration.

Runoff Coefficient

The runoff coefficient C is the variable of the Rational Method least able to be precisely determined, and its quantification requires judgment and understanding on the part of the engineer. Its use in the formula implies a fixed ration for any given drainage area. In reality this is not the case. The coefficient represents the integrated effects of infiltration, detention, evaporation, retention, flow routing, and interception which all affect the time distribution and peak rate of runoff.

Table 6.2 Rational Method Runoff Coefficients

Description of area	Runoff coefficients
Business:	
Central business areas	0.70-0.95
District and local areas	0.50-0.70
Residential:	
Single-family areas	0.35-0.45
Multiunits, detached	0.40-0.60
Multiunits, attached	0.60-0.75
Residential 1/4-hectare (1/2-acre) lots or larger	0.25-0.40
Industrial:	
Light areas	0.50-0.80
Heavy areas	0.60-0.90
Parks, cemeteries	0.10-0.25
Playgrounds	0.20-0.35
Railroad yard areas	0.20-0.40
Unimproved areas	0.10-0.30

Source: Reprinted with permission of the American Society of Civil Engineers [3].

Table 6.2 presents C values given by the American Society of Civil Engineers (ASCE) [3].

It is often desirable to develop a composite runoff coefficient based on the percentage of different types of surface in the drainage area. This procedure is often applied to typical sample subareas as a guide to selection of reasonable values of the coefficient for an entire area. Suggested coefficients with respect to surface type are given in Table 6.3. The values for streets, drives, walks, and roofs are ASCE values in Ref. 3. The values for soils are from Ref. 4. The soils classes given are those of the Soil Conservation Service (SCS) and are explained later. Type A soils are sandy with high infiltration capacity while D soils are heavy with low infiltration. These values are essentially the same as given by the ASCE.

Adjustment for Infrequent Storms

The adjustment of the Rational Method for use with major storms can be made by utilizing the right side of the rational formula by a frequency factor C_f, which is used to account for antecedent precipitation conditions. The rational formula now becomes:

$$Q = 0.275CIAC_f \quad [Q = CIAC_f] \tag{6.2}$$

Table 6.4 which shows the C_f values can be used. The product of C times C_f should not exceed 1.0.

When analyzing the major runoff occurring on an area that has a storm sewer system sized for a minor (2 to 10 year) storm, care must be used when applying the Rational Method. Normal application of the Rational Method

Table 6.3 Rational Method Runoff Coefficients for Composite Analysis

For Impervious Surfaces	
Character of surface	Runoff coefficient
Streets:	
Asphaltic	0.70-0.95
Concrete	0.80-0.95
Drives and walks	0.75-0.85
Roofs	0.75-0.95

For Pervious Surfaces					
		Runoff coefficient			
Slope		A soils	B soils	C soils	D soils
Flat:	0-2%	0.04	0.07	0.11	0.15
Average:	2-6%	0.09	0.12	0.16	0.20
Steep:	Over 6%	0.13	0.18	0.23	0.28

Sources: (Top) Reprinted with permission of the American Society of Civil Engineers [3]; (bottom) reprinted with permission of Kurt W. Bauer and The Southeastern Wisconsin Regional Planning Commission [4].

Table 6.4 Frequency Factors for Rational Formula

Recurrence interval	C_f
2-10 years	1.0
25 years	1.1
50 years	1.2
100 years	1.25

assumes that all runoff is collected by the storm sewer. In the design of a minor system, the time of concentration is dependent upon the flow time in the sewer; however, during a major runoff the sewers should be fully taxed and unable to accept all the water flowing to the inlets. The additional water then flows past the inlets and continues overland, generally at a lower velocity than the water in the storm sewers. This design requires an analysis of the split of total flow between underground flow and overland flow. Times of concentration and resulting peak discharge estimates should be evaluated for major and minor events considering this effect.

Hydrograph Approximations Using the Rational Method

The estimation of runoff hydrographs may be necessary for evaluation of very small and simple development drainage proposals, particularly when they are involved with on-site storage systems. However, such methods should not be used for larger areas or evaluation of costly structures or where serious ramifications are possible downstream.

There are many versions, but the usual concept is that a triangular hydrograph describes the response of the basin from a unit storm of duration equal to the time of concentration. The peak of the triangle (hydrograph) is equivalent to the peak discharge calculated by the rational formula. The time-to-peak of the hydrograph is equal to the time of concentration, and both are measured from the beginning of the rainfall. The receding limb of the hydrograph is also equal to the time-to-peak; thus the hydrograph is an isosceles triangle.

Application of a design storm appropriate for a watershed can be made by revising the design storm into increments which have a time equal to the time of concentration. The rainfall for each revised time increment is converted to rainfall intensity and an incremental triangular hydrograph response calculated for each. The series of triangular responses can be summed to arrive at a design hydrograph.

The reader is reminded that this method provides a rough approximation and is only appropriate for small, simple areas. A safety factor should be used appropriate to the possible variances. *One of the key variances is the amount of rainfall losses assumed by the rational coefficient C.* For major storms, and particularly in urbanized areas, the usual total percentage of rainfall that becomes runoff will be different than that indicated by the factor C. The following subsection gives insight on typical values.

In practice, one can use the rational peak discharge as indicated, but the hydrograph approximated from this method should be adjusted to more accurately reflect the probable runoff volume and other special effects.

Methods are available for evaluation of storage systems such as the FAA storage method [5] which is presented in Chap. 10 on storage and other references [6]. The key point to remember is that these are crude approximations appropriate only to small, simple facilities with no hazard potential.

6.3 DETERMINATION OF RAINFALL EXCESS

Introduction

All hydrology techniques except for the Rational Method require rainfall excess as a direct data input. As explained in the preceding chapter, this requires the estimation of rainfall losses. Several of the techniques used are presented below.

Horton's Equation [7]

Horton developed the following equation to define the rate curve of infiltration capacity:

$$f = f_c + (f_0 - f_c)e^{-kt} \tag{6.3}$$

where f = rate of infiltration (per hour)
f_c = ultimate infiltration rate (per hour)
f_0 = initial infiltration rate (per hour)
k = constant, depending primarily on soils and vegetation (typically varies from 2.0 to 10.0)
t = time from start to rainfall (hr)

The general behavior of Horton's equation is illustrated in Fig. 6.2. Values of f_0, f_c, and k are associated with soil-cover complexes and are available for certain areas. The value of f_0 is also a function of the antecedent precipitation as reflected in the initial soil moisture conditions. Practical difficulties have been encountered in estimating values of f_0 and k which are appropriate to use in specific basins.

Horton's approach provides a useful measure of infiltration behavior for design-type events where the required soil moisture/infiltration conditions can be presented.

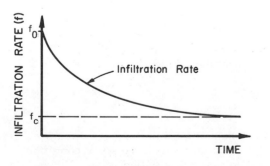

Figure 6.2 Horton's infiltration rate curve.

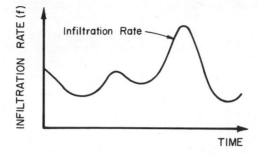

Figure 6.3 Holtan's infiltration rate curve.

Holtan's Method [8]

Holtan has developed a useful expression for the infiltration aspects of watershed modeling. This method is defined by

$$f = aF_p^{1.4} + f_c \qquad\qquad (6.4)$$

where f = infiltration capacity
$\quad\quad\;\; a$ = index of surface-connected porosity
$\quad\quad\;\; F_p$ = available storage in the soil layer(s) considered to be the "sink" for the infiltrated water
$\quad\quad\;\; f_c$ = constant rate of infiltration after prolonged wetting

The primary difference between this technique and Horton's is that the infiltration rate is a function of the present soil moisture conditions, and thus does not vary only with time, as does Horton's model; the effective infiltration rate can both increase and decay, as shown in Fig. 6.3, depending on the soil moisture conditions.

The value of F_p is computed continuously during a simulation and is affected by the inflow and outflow to the soil zone during each time increment. This may be expressed as

$$F_p^* = \hat{F}_p + (f_c - f)\,\Delta t \qquad\qquad (6.5)$$

where F_p^* = new value of F_p
$\quad\quad\;\; \hat{F}_p$ = previous value of F_p
$\quad\quad\;\; f_c$ = constant drawdown rate
$\quad\quad\;\; f$ = infiltration rate into the zone
$\quad\quad\;\; \Delta t$ = time increment

Use of this technique requires knowledge of the variables F_p, f_c, and a, as they are appropriate to the basin under study.

Soil Conservation Service Method [9-12]

Basic Approach

This technique uses three variables to estimate the rainfall excess during a given event. These variables are rainfall, the antecedent moisture condition, and the hydrologic soil cover complex. The general equation is

$$I_E = \frac{(P - I_a)^2}{P - I_a + S}$$

(6.6)

where I_E = accumulated direct runoff (rainfall excess), mm [in.]
$\quad\;\; P$ = accumulated precipitation, mm [in.]
$\quad\;\; I_a$ = initial abstraction including surface storage and infiltration
$\qquad\quad$ prior to runoff, mm [in.]
$\quad\;\; S$ = maximum potential retention, mm [in.]

Although this equation is written for cumulative rainfall and rainfall excess to any given time, a time-varying record of the rainfall excess can be easily derived.

The variable S includes I_a. An empirical relationship has been developed from data on watersheds in various parts of the United States. This generally can be expressed as

$$I_a = 0.2S$$

(6.7)

The Soil Conservation Service (SCS) has made extensive experiments and analyses of watershed data to determine the best way to relate the variable S to the soil water storage and the infiltration rates of a watershed. The method adopted is the curve number (CN) technique. This is simply a method of combining the properties of the soil groups in the watershed with both the land use and treatment classes and the antecedent moisture conditions.

The variable S in Eq. (6.6) is related to the CN by the following relationship:

$$S = \frac{1000 - 10CN}{CN}$$

(6.8)

The SCS technique is a useful and reliable method of representing the infiltration characteristics of a watershed. Once the CN is obtained, Fig. 6.4 can be used to determine the cumulative rainfall excess in millimeters or inches.

The curve number is determined by the identification and extent of various soil groups and consideration of factors such as surface conditions, vegetation cover, and other cover factors. The following paragraphs present information and guidance in the curve number selection.

The reader is encouraged to review related references for details of SCS methodology [9-12].

Determination of the Soil Group
The major soil groups are defined for the estimated watershed soil conditions. The groups, as defined by the SCS, are:

Group A: (low runoff potential): Soils having high infiltration rates even when thoroughly wetted and consisting chiefly of deep, well-drained to excessively drained sands or gravels. These soils have a high rate of water transmission.

Group B: Soils having moderate infiltration rates when thoroughly wetted and consisting chiefly of moderately deep to deep, moderately well-drained to well-drained soils with moderately fine to moderately coarse textures. These soils have a moderate rate of water transmission.

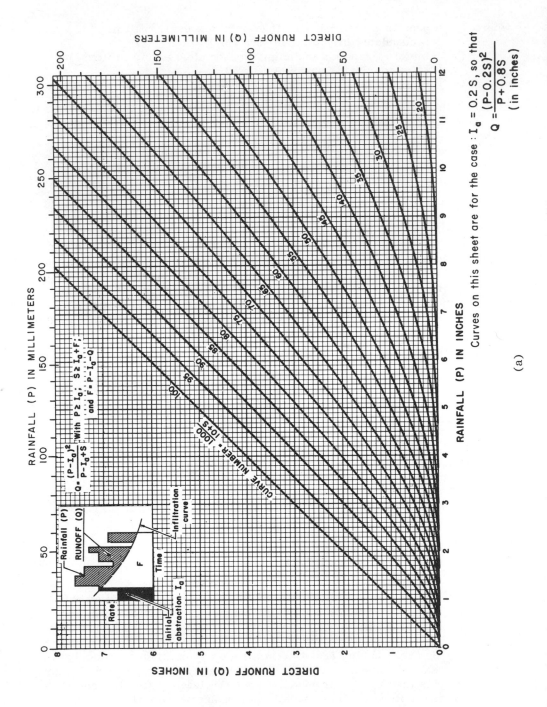

(a)

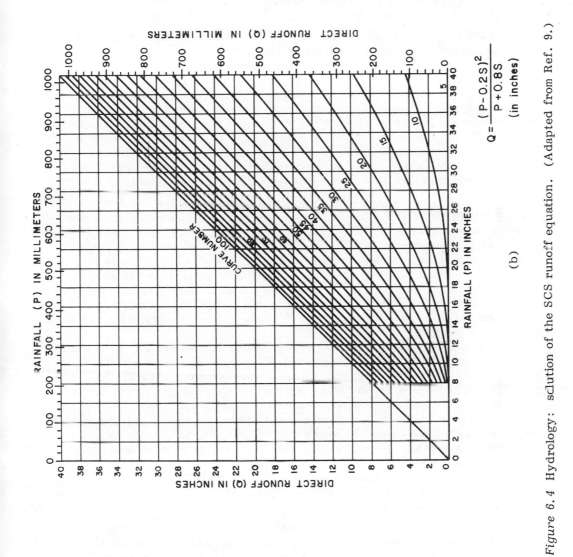

$$Q = \frac{(P - 0.2S)^2}{P + 0.8S}$$

(in inches)

(b)

Figure 6.4 Hydrology: solution of the SCS runoff equation. (Adapted from Ref. 9.)

Group C: Soils having slow infiltration rates when thoroughly wetted and consisting chiefly of soils with a layer that impedes downward movement of water, or soils with moderately fine to fine texture. These soils have a slow rate of water transmission.

Group D (high runoff potential): Soils having very slow infiltration rates when thoroughly wetted and consisting chiefly of clay soils with a high swelling potential, soils with a permanent high water table, soils with a claypan or clay layer at or near the surface, and shallow soils over nearly impervious material. These soils have a very slow rate of water transmission.

The SCS has published the Hydrologic Soil Group for the majority of the soil types found within the United States [9]. This soils list can be useful to the engineer in selecting the proper soil groups.

When determining urban CNs, consideration should be given to whether heavy equipment compacted the soil significantly more than natural conditions, whether much of the pervious area is barren with little sod established, and whether grading has mixed the surface and subsurface soils causing a completely different hydrologic condition. Any one of the above could cause a soil normally in hydrologic group A or B to be classified in group B or C. In many areas, lawns are heavily irrigated, which may significantly increase the moisture content in the soil over that under natural rainfall conditions [10].

Determination of Areal Extents and Number of Soil Groups

Precise measurement of soil group areas, such as by planimetering soil areas, is seldom necessary for hydrologic purposes. The maximum detail need not go beyond that illustrated in Fig. 6.5, where in Fig. 6.5a the individual soils in a hydrologic unit are shown on a sketch map; in Fig. 6.5b the soils are classified into groups; in Fig. 6.5c a grid (or "dot counter") is placed over the map and the number of grid intersections falling on each group is counted and tabulated; in Fig. 6.5d the tabulation and a typical computation of a group percentage is shown. Simplified versions of this procedure are generally used in practice.

Often one or two soil groups are predominant in a watershed, with others covering only a small part. Whether the small groups should be combined with those that are predominant depends on their classifications. For example, a hydrologic unit with 90 percent of its soils in the A group and 10 percent in D will have most of its storm runoff coming from the D soils. Putting all soils into the A group will cause a serious underestimation of runoff. If the groups are more nearly alike (A and B, B and C, or C and D), the under- or overestimation may not be serious, but a test may be necessary to determine this. Rather than test each case, follow the rule that two groups are combined only if one of them covers less than about 3 percent of the hydrologic unit. Impervious surfaces should always be handled separately because they produce runoff even if there are no D soils.

Determination of the Runoff Curve Number (CN)

Once the soil group is known, the runoff curve number is determined by consideration of the surface conditions, vegetation cover, and other cover factors. References 9-12 present detailed information for selection of the runoff curve number, as well as guidance toward determining a curve number for areas having mixtures of different soil and cover conditions.

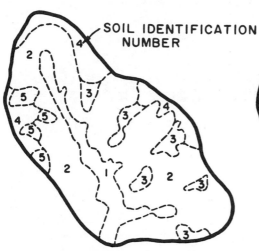

(a) DETAILED SOILS MAP   **(b) HYDROLOGIC SOIL GROUP MAP**

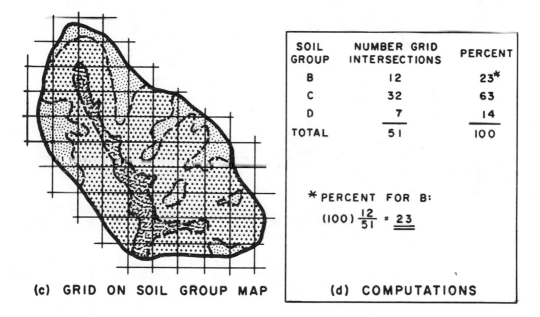

SOIL GROUP	NUMBER GRID INTERSECTIONS	PERCENT
B	12	23*
C	32	63
D	7	14
TOTAL	51	100

*PERCENT FOR B:

$(100)\dfrac{12}{51} = \underline{\underline{23}}$

(c) GRID ON SOIL GROUP MAP   **(d) COMPUTATIONS**

Figure 6.5 Steps in determining percentages of soil groups. (Courtesy U.S. Department of Agriculture, Soil Conservation Service [9].)

Table 6.5 lists CNs for agricultural, suburban, and urban land use classifications. The suburban and urban CNs are based on typical land use relationships that exist in some areas. They should only be used when it has been determined that the area under study meets the criteria for which these CNs were developed (noted in table).

Table 6.5 Runoff Curve Numbers for Selected Agricultural, Suburban, and Urban Land Use (Antecedent Moisture Condition II, and $I_a = 0.25$)

Land use description	Hydrologic soil group			
	A	B	C	D
Cultivated land:[a]				
Without conservation treatment	72	81	88	91
With conservation treatment	62	71	78	81
Pasture or range land:				
Poor condition	68	79	86	89
Good condition	39	61	74	80
Meadow: good condition	30	58	71	78
Wood or forest land:				
Thin stand, poor cover, no mulch	45	66	77	83
Good cover[b]	25	55	70	77
Open spaces, lawns, parks, golf courses, cemeteries, etc. :				
Good condition: grass cover on 75% or more of the area	39	61	74	80
Fair condition: grass cover on 50% to 75% of the area	49	69	79	84

	Average % Impervious[d]				
Commercial and business areas (85% impervious)		89	92	94	95
Industrial districts (72% impervious)		81	88	91	93
Residential:[c]					
Average lot size					
0.05 hectare (1/8 acre) or less	65	77	85	90	92
0.10 hectare (1/4 acre)	38	61	75	83	87
0.13 hectare (1/3 acre)	30	57	72	81	86
0.20 hectare (1/2 acre)	25	54	70	80	85
0.90 hectare (1 acre)	20	51	68	79	84
Paved parking lots, roofs, driveways, etc.[e]		98	98	98	98
Streets and roads:					
Paved with curbs and storm sewers[e]		98	98	98	98
Gravel		76	85	89	91
Dirt		72	82	87	89

[a]For a more detailed description of agricultural land use curve numbers refer to Ref. 9, Chap. 9.

[b]Good cover is protected from grazing and litter and brush cover soil.

[c]Curve numbers are computed assuming the runoff from the house and driveway is directed toward the street with a minimum of roof water directed to lawns where additional infiltration could occur.

[d]The remaining pervious areas (lawn) are considered to be in good pasture condition for these curve numbers.

[e]In some warmer climates of the country a curve number of 95 may be used.

Source: U.S. Department of Agriculture, Soil Conservation Service [9].

There will be areas to which the values in Table 6.5 do not apply. The percentage of impervious area for the various types of residential areas or the land use condition for the pervious portions may vary from the conditions assumed. A curve for each pervious CN can be developed to determine the composite CN for any density of impervious area. Figure 6.6 has been developed assuming a CN of 98 for the impervious area. The curves in Fig. 6.6 can help in estimating the increase in runoff as more and more land within a given area is covered with impervious material.

Chapter 4 of *Forest and Range Hydrology Handbook* [11] describes how CN is determined for national and commercial forests in the Eastern United States. Section 1 of *Handbook on Methods of Hydrologic Analysis* [13] describes how CN is determined for forest range regions in the Western United States.

In the forest range regions of the Western United States, soil group, cover type, and cover density are the principal factors used in estimating CN. Figure 6.7 shows the relationship between these factors and CN for soil-cover complexes used to date. Similar Australian forest covers are indicated for general interest.

Curve Number Modification for Antecedent Moisture and Special Conditions

Retention parameters can be modified for various situations and to allow for antecedent conditions. A key example of the need for modification is when major storage facilities are planned that would be adversely affected by antecedent conditions.

The SCS runoff equation states that 20 percent of the potential maximum retention S is the initial abstraction I_a, which is the interception, infiltration, and surface storage occurring before runoff begins. The remaining 80 percent is mainly the infiltration occurring after runoff begins. This later infiltration is controlled by the rate of infiltration at the soil surface, or by the

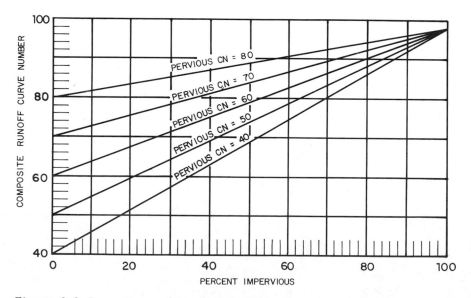

Figure 6.6 Percentage of impervious areas vs. composite CNs for given area CNs.

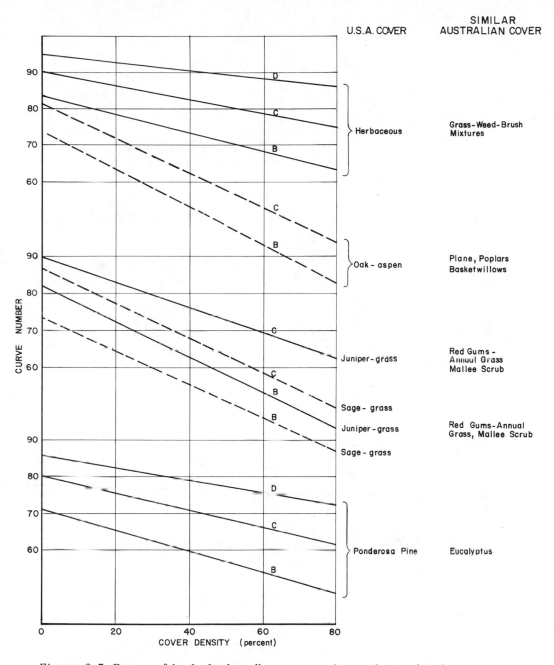

Figure 6.7 Range of hydrologic soil cover complex and associated curve numbers.

rate of transmission in the soil profile, or by the water storage capacity of the profile, whichever is the limiting factor. A succession of storms, such as one a day for a week, reduces the magnitude of S each day because the limiting factor does not have the opportunity to completely recover its rate or capacity

Table 6.6 Curve Numbers and Constants ($I_a = 0.25$)

CN for condition II (1)	CN for conditions I (2)	CN for conditions III (3)	S values[a] (in.) (4)	Curve[a] starts where P = (in.) (5)	CN for condition II (1)	CN for conditions I (2)	CN for conditions III (3)	S values[a] (in.) (4)	Curve[a] starts where P = (in.) (5)
100	100	100	0	0	60	40	78	6.67	1.33
99	97	100	0.101	0.02	59	39	77	6.95	1.39
98	94	99	0.204	0.04	58	38	76	7.24	1.45
97	91	99	0.309	0.06	57	37	75	7.54	1.51
96	89	99	0.417	0.08	56	36	75	7.86	1.57
95	87	98	0.526	0.11	55	35	74	8.18	1.64
94	85	98	0.638	0.13	54	34	73	8.52	1.70
93	83	98	0.753	0.15	53	33	72	8.87	1.77
92	81	97	0.870	0.17	52	32	71	9.23	1.85
91	80	97	0.989	0.20	51	31	70	9.61	1.92
90	78	96	1.11	0.22	50	31	70	10.0	2.00
89	76	96	1.24	0.25	49	30	69	10.4	2.08
88	75	95	1.36	0.27	48	29	68	10.8	2.16
87	73	95	1.49	0.30	47	28	67	11.3	2.26
86	72	94	1.63	0.33	46	27	66	11.7	2.34
85	70	94	1.76	0.35	45	26	65	12.2	2.44
84	68	93	1.90	0.38	44	25	64	12.7	2.54
83	67	93	2.05	0.41	43	25	63	13.2	2.64

1	2	3	4	5
82	66	92	2.20	0.44
81	64	92	2.34	0.47
80	63	91	2.50	0.50
79	62	91	2.66	0.53
78	60	90	2.82	0.56
77	59	89	2.99	0.60
76	58	89	3.16	0.63
75	57	88	3.33	0.67
74	55	88	3.51	0.70
73	54	87	3.70	0.74
72	53	86	3.89	0.78
71	52	86	4.08	0.82
70	51	85	4.28	0.86
69	50	84	4.49	0.90
68	48	84	4.70	0.94
67	47	83	4.92	0.98
66	46	82	5.15	1.03
65	45	82	5.38	1.08
64	44	81	5.62	1.12
63	43	80	5.87	1.17
62	42	79	6.13	1.23
61	41	78	6.39	1.28
42	24	62	13.8	2.76
41	23	61	14.4	2.88
40	22	60	15.0	3.00
39	21	59	15.6	3.12
38	21	58	16.3	3.26
37	20	57	17.0	3.40
36	19	56	17.8	3.56
35	18	55	18.6	3.72
34	18	54	19.4	3.88
33	17	53	20.3	4.06
32	16	52	21.2	4.24
31	16	51	22.2	4.44
30	15	50	23.3	4.66
25	12	43	30.0	6.00
20	9	37	40.0	8.00
15	6	30	56.7	11.34
10	4	22	90.0	18.00
5	2	13	190.0	38.00
0	0	0	Infinity	Infinity

[a]For CN in column 1.

Source: U.S. Department of Agriculture, Soil Conservation Service [9].

through weathering, evapotranspiration, or drainage; but there is enough
recovery, depending on the soil-cover complex, to limit the reduction.
During such a storm period, the magnitude of S remains virtually the same
after the second or third day even if the rains are large so that there is,
from a practical viewpoint, a lower limit of S for a given soil-cover complex.
Similarly, there is a practical upper limit to S, again depending on the soil-
cover complex, beyond which the recovery cannot take S unless the complex
is altered.

In the SCS method, the change in S (actually in CN) is based on an
antecedent moisture condition (AMC) determined by the total rainfall in the
5-day period preceding a storm. Three levels of AMC are used: AMC-I is
the lower limit of moisture or the upper limit of S, AMC-II is the average condition for which the CNs of Table 6.6 apply, and AMC-III is the upper limit of
moisture or the lower limit of S. The CNs for high and low moisture levels
were empirically related to the average CNs (AMC-II) of Table 6.6. Comparisons of computed and actual runoffs show that for most problems the extreme
AMC can be ignored and the average of CN of Table 6.6 applied [9].

Other Example Techniques

Richard's Equation
A rigorous analysis of soil moisture movement based on Darcy's law and the
equation of continuity leads to the second-order nonlinear partial differential
equation:

$$\frac{\partial \theta}{\partial t} = - \frac{\partial}{\partial Z} K(\theta) \left(\frac{\partial \psi(\theta)}{\partial Z} + 1 \right) \tag{6.9}$$

where θ = soil moisture content
$K(\theta)$ = hydraulic conductivity (a function of moisture content)
$\psi(\theta)$ = capillary suction (a function of moisture content)
t = time
Z = depth below the surface

Equation (6.9), often referred to as Richard's equation or the diffusion
equation, may be solved by finite difference techniques for soil moisture
movement through the surface (i.e., infiltration) and yields excellent agreement with laboratory experiments. However, the method is difficult to apply
and requires knowledge of the $K(\theta)$ and $\psi(\theta)$ functions, the latter of which
exhibits significant hysteresis for most soils. The prohibitive amount of information required by Richard's equation forces engineers to look to simpler
methods which may be based on simplified theories or derived empirically from
field results of runoff studies. In natural catchments the situation is complex
because of the variety and combinations of soil characteristics, soil moisture
conditions, and other factors occurring in nature.

Philip's Two-Parameter Equation [14, 15]
Philip developed an approximation to the solution of Richard's equation for a
homogeneous soil in the form of an infinite series; because of rapid convergence, only the first two terms of the series need to be considered, giving:

$$F = St^{1/2} + f_c t \tag{6.10}$$

where f_c = ultimate infiltration rate
F = infiltration volume
S = soil sorptivity
t = time

The value of f_c is associated with the particular soil-cover complex and may be identified with the similar parameter found in other formulations. S is dependent on the soil and its initial moisture content. Practical difficulties in estimating values of S for use in specific basins and with historic events have been encountered, although in computer modeling approaches where soil moisture is accounted, S can easily be related to soil moisture storage levels.

For design-type events, the two-parameter Philip equation provides a useful description of infiltration with physical significance. The first term on the right-hand side of the equation represents the time-dependent component of infiltration due to capillary suction effects, while the second term represents the component due to gravity.

Green and Ampt Method [16]
This method employs the fact that an abrupt wetting front forms in a soil during infiltration. When moisture is infiltrating at the capacity rate, soil above the wetting front is at saturation and soil below the wetting front is essentially at the initial moisture content. By simplifying the soil moisture profile as indicated in Fig. 6.8 and applying Darcy's law to the wetted zone, the Green and Ampt equation is derived:

$$f = K_s \frac{L + \psi}{L} \tag{6.11}$$

where f = infiltration rate
K_s = saturated conductivity of the soil
L = depth of the wetted zone
ψ = capillary suction at the wetting front

Equation (6.11) is not in itself immediately useful, but when applied to the simplified profile at the precise moment when saturation occurs at the surface, which is the moment when surface runoff commences, the following equation can be derived:

$$F = \psi \frac{d}{I/K_s - 1} \tag{6.12}$$

where F = infiltration volume
I = rainfall intensity
d = relative moisture deficit when rainfall commences

that is,

$$d = 1.0 - \frac{\text{initial moisture content}}{\text{saturated moisture content}} \tag{6.13}$$

Providing only that rainfall intensity prior to commencement of runoff may reasonably be regarded as a constant or that an average value of 1 is used, Eq. (6.12) allows the computation of the volume of infiltration (and hence the time) to commencement of surface runoff.

Once surface runoff has commenced, Eq. (6.11) is directly applicable, but a more useful formulation of it is as follows:

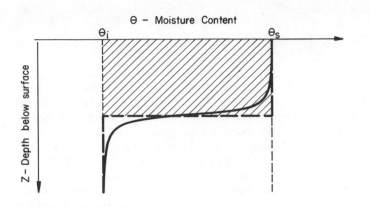

Figure 6.8 A Typical moisture content profile during infiltration. The continuous heavy line shows the true profile; the dashed heavy line is the simplified profile; θ_i = initial moisture content; θ_s = saturated moisture content; shaded area represents the current volume of infiltration.

$$f = K_s \left(1 + \psi \frac{d}{F}\right) \qquad\qquad (6.14)$$

This equation allows infiltration rate f (and hence volume of infiltration F) to be progressively computed as the storm proceeds. An iterative procedure is necessary for solution of Eq. (6.14), but convergence is rapid.

To recapitulate, Eq. (6.12) is used to determine the time at which surface runoff commences, while Eq. (6.14) is used to determine subsequent infiltration rates.

One advantage of the Green and Ampt Method is that the initial moisture deficit appears explicitly in the equations. K_s is a parameter with a clear physical meaning and so the possibility exists of evaluating it through experiment or field tests. ψ is the average capillary suction at the wetting front, and because of the shape of the $\psi\theta$ function for most soils during wetting, it often remains constant regardless of the initial moisture content of the soil. In any case, use of the method with ψ assumed dependent only on the soil-cover complex has yielded very good agreement with both Richard's equation and experimental results. The physical significance of the two parameters K_s and ψ may be readily appreciated and related by the engineer to his personal knowledge of the soils in the catchment being modeled.

Table 6.7 presents an example computation. Note that the rainfall data is in 12-min increments where the first 12-min increment has a rainfall of 6.8 mm or the equivalent intensity I of 34.0 mm/hr. Using this value an initial infiltration of 3.4 mm has to be satisfied according to Eq. (6.12). This would occur during the first 6 min at the above rainfall rate. Thus, the iterative computation with Eq. (6.14) begins after 6 min. An infiltration rate is assumed which leads to an assumed cumulative infiltration (column 6). This value is used to calculate the infiltration rate f in Eq. (6.14). When the assumed and calculated infiltration rates correspond, this value should be used. This infiltration is then deducted from the rainfall and, further, any detention losses are subtracted. The calculation then proceeds until infiltration continually exceeds rainfall.

Simple Abstraction Method

A common method of accounting for infiltration is the Simple Abstraction Method. In this method, the initial infiltration is estimated as a fixed amount in millimeters and the steady state, or ultimate rate of infiltration, is estimated in millimeters per hour. Typical values are given in the previous chapter.

The initial infiltration f_0 is satisfied in full up to the amount of the rainfall in the first rainfall increment with the balance being carried to the second increment of rainfall and then to the third until the total f_0 is fulfilled. Subsequently, the steady-state infiltration rate in millimeters per hour is satisfied up to the amount of the rainfall intensity, which is also expressed in millimeters per hour for the rainfall period increment.

Often the initial infiltration is lumped together with the depression and detention losses for computational simplicity.

Rainfall Excess Example

That portion of rainfall that becomes runoff during or soon after a storm is called *rainfall excess*. The abstractions from rainfall that determine the rainfall excess are functions of infiltration, detention and depression storage, intensity of rainfall, percent of imperviousness, etc.

An example of estimating the rainfall excess is presented below and tabulated in Table 6.8.

Column 1: For the design location select a rainfall time interval, usually 6 or 12 min for small urban basins, 12 to 36 min for large basins.

Column 2: Determine the incremental precipitation for the design storm of interest from Chap. 4 on rainfall.

Column 3: Add the incremental precipitation to provide a cumulative tabulation for the storm time period.

Pervious Area

(Columns 4 to 8): The example utilizes Horton's equation, but similar tabulations would be utilized for Holtan's and other methods. Columns 4 and 5 would not be necessary for the SCS method since the cumulative rainfall excess is determined directly.

Column 4: A total detention value is assumed in accordance with Table 5.2 (Chap. 5). This value is then allocated against the first time increments of rainfall until the total detention value is satisfied.

Column 5: The infiltration function chosen is used to assign potential infiltration demands midway in a time period. In the example, potential infiltration demands are calculated at 6, 18, 30 min, and so on.

Column 6: The actual infiltration is assigned by comparing the rainfall remaining after detention losses. If the remaining rainfall for a time increment is less than the infiltration demand, then the value of the incremental rainfall is assumed to be equal to the actual infiltration rate.

Column 7: The rainfall excess is then equal to column 2 minus column 4 minus column 6; however, this column is determined directly in the SCS method.

Column 8: Column 7 times the (decimal) percent of the pervious area gives the area-weighted depth of water that will run off in each time increment for the pervious area.

Table 6.7 Rainfall Excess Computation Example Using Green and Ampt Method

Time (min) (1)	Incremental precipitation (mm) (2)	Cumulative precipitation (mm) (3)	Assumed infiltration rate (mm/hr) (4)	Incremental infiltration (mm) (5)	Cumulative infiltration (mm) (6)	Calculated infiltration rate Eq. (6.14) (7)	Detention depression (mm) (8)	Rainfall excess (mm) (9) (2) − (5) − (8)
0	0	0	—	—	0		0	0
6.0[a]	3.4	3.4		3.4	3.4		0	0
12	3.4	6.8	(i) 23.1	2.3	5.7	27.3	0	0
			(ii) 26.0 Use	2.6	6.0	26.0	0.8	0
24	6.8	13.6	(i) —	5.2	11.2			
			(ii) 15.0	3.0	9.0	17.7		
			(iii) 16.0	3.2	9.2	17.3		
			(iv) 17.0 Use	3.4	9.4	17.0	3.4	0
36	12.2	25.8	(i) —	3.2	12.8			
			(ii) 12.8	2.6	12.0	13.5		
			(iii) 13.5 Use	2.7	12.1	13.4	5.8	3.7
48	12.2	38.0	(i) —	2.7	14.8			
			(ii) 11.0	2.2	14.3	11.5	0	
			(iii) 11.4 Use	2.3	14.4	11.4	0	9.9

60	3.6	41.6	(i)	—		2.3	16.7			
			(ii)	10.1		2.0	16.4	10.1	1.6	0
72	3.6	45.2	(i)	—		2.0	18.4			
			(ii)	9.0	Use	1.8	18.2	9.2	1.8	0
84	2.2	47.4	(i)	—		1.8	20.0			
			(ii)	8.5	Use	1.7	19.9	8.5	0.5	0
96	2.0	49.4	(i)	—		1.7	21.6			
			(ii)	8.0	Use	1.6	21.5	8.0	0.4	0
108	1.4	50.8	(i)	—		1.6	23.1			
			(ii)	7.5		1.5	23.0	7.5		
					Actual	1.4	22.9		0	0
120	1.2	52.0	(i)	—		1.5	24.4			
			(ii)	7.0		1.4	24.3	7.2		
			(ii)		Actual	1.2	24.1		0	0

Summary: 52 mm = 24.1 mm infiltrated
            + 10.0 mm depression/retention
            + 17.9 mm rainfall excess

ψ = 500 mm
K_s = 1.0 mm/hr
d = 0.3

[a]Initial rainfall rate I = 34.0 mm/hr, giving F = 3.4 mm by Eq. (6.12); at rate of 34 mm/hr this volume will be satisfied in 3.4 mm /(34 mm/hr) = 0.1 hr = 5 min.

Table 6.8 Determination of Rainfall Excess, Typical Melbourne Creek, Australia—Design Storm: 100-Year, 2-Hr (Summer)

| | | | | Pervious | |
| | | | | --- | --- |
Time (min) (1)	Incremental precipitation (mm) (2)	Cumulative precipitation (mm) (3)	Detention depression (mm) (4)	Infiltration function[a] (mm) (5)	Actual infiltration (mm) (6)
0	0	0	0	–	0
12	6.8	6.8	6.8	10.5	0
24	6.8	13.6	3.2	7.2	3.6
36	12.2	25.8	0	4.9	4.9
48	12.2	38.0	0	3.4	3.4
60	3.6	41.6	0	2.4	2.4
72	3.6	45.2	0	1.8	1.8
84	2.2	47.4	0	1.3	1.3
96	2.0	49.4	0	1.0	1.0
108	1.4	50.8	0	0.8	0.8
120	1.2	52.0	0	0.7	0.7
			10		19.9

[a]Horton's infiltration function where

f_0 = 63.5 mm/hr
f_c = 2.0 mm/hr
K = 2

General guidance for selection of K values is given in Ref. 17. Experience with infiltration data and field charactertistics of various soils, surface conditions, and vegetation cover allows selection of appropriate K values.

Impervious Area (columns 9 to 12)

Column 9: Enter the total assumed impervious detention and depression storage, determined from Table 5.2, at the bottom of column 9. The impervious detention and depression storage in column 9 is then either the amount of precipitation in column 2 or the amount available as determined by deducting the total accumulated amount from the total assumed value shown at the bottom of column 9. When the total assumed amount is fully used, all remaining values are zero.

Column 10: Rainfall excess for the impervious area is column 2 less column 9.

Column 11: Column 10 times the (decimal) percent of impervious area gives the area-weighted depth of water that will run off in each time increment for the pervious area.

area (75%)		Impervious area (25%)			
Excess Precipitation (mm) (7)	75% Excess precipitation (mm) (8)	Detention depression (mm) (9)	Excess precipitation (mm) (10)	25% Excess precipitation (mm) (11)	Total average excess precipitation (mm) (12)
0	0	0	0	0	0
0	0	2.5	4.3	1.1	1.1
0	0	0	6.8	1.7	1.7
7.3	5.5	0	12.2	3.1	8.6
8.8	6.6	0	12.2	3.1	9.7
1.2	0.9	0	3.6	0.9	1.8
1.8	1.4	0	3.6	0.9	2.3
0.9	0.7	0	2.2	0.6	1.3
1.0	0.8	0	2.0	0.5	1.3
0.6	0.4	0	1.4	0.4	0.8
0.5	0.4	0	1.2	0.3	0.7
22.1	16.7	2.5	49.5	12.6	29.3

Total Area

> *Column 12*: Add column 11 and column 8 to obtain the average rainfall excess. This is the "design rainfall excess" that will be applied to obtain storm runoff hydrographs.

6.4 SYNTHETIC UNIT HYDROGRAPH PROCEDURE (SUHP)

Introduction

The Synthetic Unit Hydrograph Procedure (SUHP) begins with the derivation of a unit hydrograph for the basin of interest. This unit hydrograph is then used to estimate the runoff hydrograph according to the design rainfall.

General Equations and Relationships

Snyder proposed two basic equations to be used in defining the synthetic unit hydrograph [18]. The first equation defines the lag time of the basin in terms of time-to-peak t_p which, for the SUHP, is defined as the time from the center of the unit storm duration to the peak of the unit hydrograph:

$$t_p = C_t(0.3861LL_{ca})^{0.3} \qquad \left[t_p = C_t(LL_{ca})^{0.3} \right] \tag{6.15}$$

where t_p = time-to-peak hydrograph from midpoint of unit rainfall, hr
L = length along stream from study point to upstream limits of the basin, km [mi]
L_{ca} = distance from study point along stream to the centroid to the basin, km [mi]
C_t = a coefficient reflecting time to peak

The second equation defines the unit peak of the unit hydrograph:

$$q_p = \frac{7.0C_p}{t_p} \qquad \left[q_p = \frac{640C_p}{t_p} \right] \tag{6.16}$$

where q_p = peak rate of runoff, $m^3/(sec)(km^2)$ $[ft^3/(sec)(mi^2)]$
C_p = a coefficient related to peak rate of runoff

Victor Mockus [9] derived a dimensionless unit hydrograph which is presented in Table 6.9. It can provide general shaping information and is useful in deriving other basic relationships helpful to the Synthetic Unit

Table 6.9 Ratios for Dimensionless Unit Hydrograph

Time ratios (t/T_p)	Discharge ratios (q/q_p)	Time Ratios (t/T_p)	Discharge ratios (q/q_p)
0	0.000	1.6	0.560
0.1	0.030	1.7	0.460
0.2	0.100	1.8	0.390
0.3	0.190	1.9	0.330
0.4	0.310	2.0	0.280
0.5	0.470	2.2	0.207
0.6	0.660	2.4	0.147
0.7	0.820	2.6	0.107
0.8	0.930	2.8	0.077
0.9	0.990	3.0	0.055
1.0	1.000	3.2	0.040
1.1	0.990	3.4	0.029
1.2	0.930	3.6	0.021
1.3	0.860	3.8	0.015
1.4	0.780	4.0	0.011
1.5	0.680	4.5	0.005

Source: U.S. Department of Agriculture, Soil Conservation Service [9].

Hydrograph Procedure; however, it should not be used directly for representing urban areas.

The dimensionless curvilinear unit hydrograph in Fig. 6.9 has 37.5 percent of the total volume in the rising side, which is represented by 1 unit of time and 1 unit of discharge. This dimensionless unit hydrograph also can be represented by an equivalent triangular hydrograph having the same units of time and discharge, thus having the same percent of volume in the rising side of the triangle.

The base of the approximate triangle in Fig. 6.9 can be derived with the following results:

$$T_b = \frac{1.00}{0.375} = 2.67 \text{ units of time or } 2.67 T_p \tag{6.17}$$

$$T_r = T_b - T_p = 1.67 \text{ units of time or } 1.67 T_p \tag{6.18}$$

These relationships lead to the following equation:

$$q_p = \frac{5.29}{T_p} \quad \left[q_p = \frac{484}{T_p} \right] \tag{6.19}$$

where

$$T_p = \frac{t_u}{2} + t_p \tag{6.20}$$

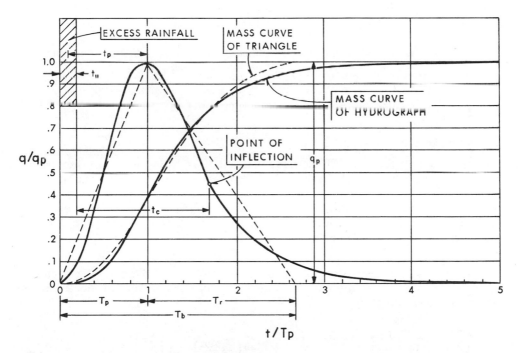

Figure 6.9 SCS dimensionless curvilinear unit hydrograph and equivalent triangular hydrograph. (Reproduced from Ref. 9.)

However, SCS notes that the numerator of 5.29 [484] is known to vary from 6.56 [600] to 3.28 [300], similar to variances noted by Snyder [9, 18]. Further, a relationship is derived to define the recommended unit rainfall duration for the unit hydrograph as:

$$t_u = 0.133t_c \quad \text{or} \quad \frac{t_c}{7.5} \tag{6.21}$$

where t_c = time of concentration as illustrated in Fig. 6.9 and the average relationship between the lag time t_p and the time of concentration is:

$$t_p = 0.6t_c \tag{6.22}$$

General Guidelines for Determining Parameters

The previous discussion presents basic relationships of synthetic unit hydrograph theory. Final determination of a synthetic unit hydrograph for a given watershed requires additional information. For example, values of C_t in Eq. (6.15) are needed to determine the lag time. The basic approach is to determine these types of coefficients or equations by analysis of rainfall runoff data from other similar watersheds. The HEC-1 computer program [28] provides an efficient means of analyzing such data.

Fortunately, many other hydrologists have already performed such analyses of local, regional, and national data and have arrived at coefficients and variations of and additions to the basic unit hydrograph equations. Any of these relationships, which are presented below, have potential application to the basin being studied and thus provide a savings in study effort. But the hydrologist must determine which equations are applicable based upon the similarity between the study basin and the watersheds for which the equations were derived.

The relationships available can generally be put into three categories:

1. Estimation of lag time t_p
2. Estimation of a synthetic unit hydrograph discharge peak q_p
3. Unit hydrograph shape factors

Estimation of Lag Time t_p

The various unit hydrograph methods are sensitive to lag time or other peak response time factors. The unit hydrograph peak discharge and shaping factors are usually a function of the lag time and other parameters. Most methods are usually derived with algorithms that are a direct or indirect function of the lag time. Thus, the determination of the lag time is critical to the reliability of the results.

Rural Areas

The Tulsa District Army Corps of Engineers [19] has derived a relationship for t_p based upon data for natural watersheds in the central and northeastern Oklahoma area, which is:

$$t_p = 0.1842\left(\frac{LL_{ca}}{\sqrt{S}}\right)^{0.39} \quad \left[t_p = 1.42\left(\frac{LL_{ca}}{\sqrt{S}}\right)^{0.39}\right] \tag{6.23}$$

where S   = watershed slope, m/m [ft/mi]
     L   = stream length, km [mi]
    L_{ca} = length along stream to centroid of basin, km [mi]

This equation, illustrated in Fig. 6.10, is recommended for natural watersheds. Note that the data and derivation were performed originally in British units; thus the graph is presented in these units only. It can be checked for unusual cases by estimating the time of concentration and multiplying by a factor of 0.60 as indicated earlier.

Also shown in Fig. 6.10 are relationships for California mountain and foothill regions [20, 26] as follows:

California mountains:

$$t_p = 0.1642\left(\frac{LL_{ca}}{\sqrt{S}}\right)^{0.38} \qquad \left[t_p = 1.2\left(\frac{LL_{ca}}{\sqrt{S}}\right)^{0.38}\right] \qquad (6.24)$$

California foothills:

$$t_p = 0.0985\left(\frac{LL_{ca}}{\sqrt{S}}\right)^{0.38} \qquad \left[t_p = 0.72\left(\frac{LL_{ca}}{\sqrt{S}}\right)^{0.38}\right] \qquad (6.25)$$

A report by Espey, et al. [21] for watersheds in Texas, New Mexico, and Oklahoma resulted in the following recommended equation:

$$T_R = 3.056 L_f^{0.12} S_f^{-0.52} \qquad \left[T_R = 2.65\, L_f^{0.12} S_f^{-0.52}\right] \qquad (6.26)$$

where T_R = time of rise in minutes which can be assumed to be equal to

$$T_R = T_p = \frac{t_u}{2} + t_p \qquad (6.27)$$

and L_f = stream length, m [ft]
    S_f = slope, m/m [ft/ft]

This equation is based on data from small watersheds ranging as follows: L, 990 to 7700 m [3250 to 25,300 ft]; S, 0.008 to 0.015 m/m [ft/ft]; and T_R, 30 to 150 min.

Urbanized Areas
Several approaches are presented here which should be used with judgment and in comparison to arrive at a recommended t_p.

The Tulsa District Army Corps has derived parallel relationships for 50 and 100 percent urbanized basins as follows:

For 50 percent urbanized:

$$t_p = 0.1193\left(\frac{LL_{ca}}{\sqrt{S}}\right)^{0.39} \qquad \left[t_p = 0.92\left(\frac{LL_{ca}}{\sqrt{S}}\right)^{0.39}\right] \qquad (6.28)$$

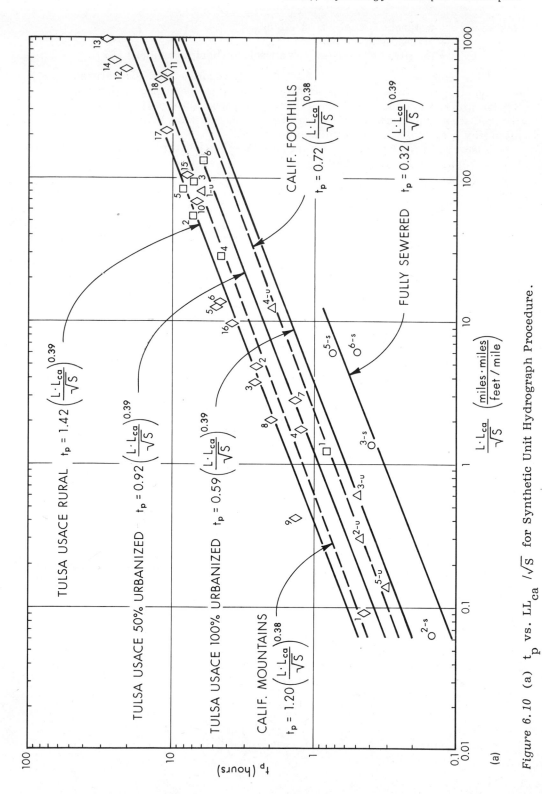

Figure 6.10 (a) t_p vs. $LL_{ca}/\sqrt{S}$ for Synthetic Unit Hydrograph Procedure.

OKLAHOMA, RURAL (by Tulsa, USACE) Ref. **19**

		A sq. mile	S ft/ mile	L miles	L_ca miles
◇1	Little Dry Ck., Alex, OK	.88	82.1	1.4	0.6
◇2	Worley Ck., Tuttle, OK	11.2	17.4	6.3	3.2
◇3	West Beaver, Orland, OK	13.9	23.8	6.4	2.9
◇4	Canyon View, Geary, OK	11.8	19.4	4.4	1.9
◇5	Dry Creek, Kendrick, OK	69	10.9	10.7	3.9
◇6	Elm Creek, Foraker, OK	18.2	17.5	9.4	5.8
◇7	Rock Creek, Snider, OK	9.1	35.8	5.4	3.1
◇8	Adams Ck., Beggs, OK	5.9	32.1	4.4	2.7
◇9	Corral Ck., Yale, OK	2.89	53.3	2.4	1.3
◇10	Big Hill Ck., Cherryvale, KS	37	11.1	21.8	10.6
◇11	Bird Creek, Avant, OK	364	5.8	52.4	25.8
◇12	L. Caney R., Copan(Upper), OK	424	5.1	50.4	26.7
◇13	L. Caney R., Copan(Lower), OK	502	4.1	60.5	33.0
◇14	Hominy Ok., Shiatook, OK	340	5.5	55.2	29.0
◇15	Polecat Ck., Heyburn Dam	133	9.0	25.8	12.2
◇16	Council Ck., Stillwater, OK	31	12.1	8.6	4.0
◇17	Pryor Ck., Pryor, OK	229	5.3	36	14
◇18	Sand Ck., Okesa, OK	139	13.5	60	31

OKLAHOMA, URBAN (by Tulsa, USACE) Ref. **19**

			A sq. mile	S ft/ mile	L miles	L_ca miles
⩟1-u	Deep Fork R., Arcadia, OK.	(30% Urbanized)	108	10.3	25.8	10.0
△2-u	Bluff Ck., Okla. City, OK	(60% ")	1.64	62.7	2.18	1.14
△3-u	Deep Fork Ck., Okla. City, OK.	(100% ")	2.98	44.9	2.88	1.44
△4-u	Deep Fork Ck., Eastern Ave. Okla City, OK	(100% ")	20.3	19.2	11.4	4.8
△5-u	Crutch Ck., Trib, Okla. City, OK	(60% ")	0.47	49.1	1.45	0.7

TEXAS AND ILLINOIS, URBAN Ref. **21**

		I per- cent	A sq. mile	S ft/ mile	L miles	L_ca miles	COMMENTS
□1	Boneyard, Illinois	37.4	4.45	9.5	2.0	1.3	storm sewers, no channel improvements
□2	Brays Bayour, Houston, TX	40.0	88.4	4.07	23.3	10.4	storm sewers and channel improvements
□3	Greens Bayou, Houston, TX	25.0	67.5	6.65	21.6	10.0	agricultural & urban, no storm sewers
□4	Halls Bayou, Houston, TX	30.0	26.2	7.08	13.5	5.7	some storm sewers & channel improvements
□5	Simms Bayou, Houston, TX	30.0	63.0	3.38	18.0	9.7	some storm sewers & channel improvements
□6	White Oak Bayou, Houston, TX	35.0	92.0	5.02	21.1	12.8	storm sewers & channel improvements

KENTUCKY, FULLY SEWERED
(corrected data from Ref. **22**)

		I per- cent	A sq. mile	S* ft/ mile	L miles	L_ca miles
⊙2-s	17th St., Louisville, KY	83	.22	20.06	.93	.31
⊙3-s	NW Trunk, Louisville, KY	50	1.90	6.34	3.03	1.13
⊙5-s	Southern Outfall, Louisville KY	48	6.44	7.23	6.44	2.52
⊙6-s	S.W. Outfall, Louisville, KY	33	7.51	7.76	6.48	2.68

*Slope used is the Weighted Sewer Slope

(b)

Figure 6.10 (b) Explanation of data points.

For 100 percent urbanized:

$$t_p = 0.07653 \left(\frac{LL_{ca}}{\sqrt{S}} \right)^{0.39} \quad \left[t_p = 0.59 \left(\frac{LL_{ca}}{\sqrt{S}} \right)^{0.39} \right] \tag{6.29}$$

as defined with the same units as Eq. (6.15).

The Army Corps estimates that the 100 percent urbanized basin would have approximately 50 percent impervious area. The Army Corps indicates that these relationships are questionable in extremely large basins and in small basins less than 1/3 to 1/2 mi^2 and that these are based on a limited data base. Thus, the information should be used carefully.

Also plotted are urban data by Van Sickle recorded in Ref. 21.

Interestingly, Eagleson presents synthetic unit hydrograph data of fully storm-sewered basins [22]. When corrected for slope defined as ft/mi, these data can be plotted on Fig. 6.10 and a parallel relationship drawn to the Army Corps'. The plotted line has the relationship:

$$t_p = 0.0415 \left(\frac{LL_{ca}}{\sqrt{S}} \right)^{0.39} \quad \left[t_p = 0.32 \left(\frac{LL_{ca}}{\sqrt{S}} \right)^{0.39} \right] \tag{6.30}$$

However, in this equation the slope is the weighted slope of the storm sewers. Since the Tulsa District Army Corps data were based on a mixture of basins that carried mainstream flows in open channels, either artificial or natural waterways, all the various relationships are deemed reasonable and usable. Also, usage for small watersheds is reasonable.

However, another precaution is noted by the Army Corps. That is that the routing effects of the mainstreams and the associated valley storage should be evaluated. It may be necessary to derive hydrographs for individual subbasins and route the discharges through a main channel. Such streamflow routing is discussed later.

Epsey et al. [21] proposed the following relationship in urban areas for the time from the beginning of effective rainfall to the peaks:

$$T_R = 29.36 UL_f^{0.29} S_f^{-0.11} I^{-0.61} \quad \left[T_R = 20.8 UL_f^{0.29} S_f^{-0.11} I^{-0.61} \right] \tag{6.31}$$

where U = 1.0 for natural conditions
       = 0.8 for watersheds with some storm sewers and channelization
       = 0.6 for watersheds with extensive urban development
    L_f = stream length m [ft]
    S_f = slope, m/m [ft/ft]
     I = percent impervious

This equation is based on data from watersheds ranging as follows: L, 60 to 16,700 m [200 to 54,800 ft]; S, 0.0064 to 0.0104 m/m [ft/ft]; I, 2.7 to 100 percent; and T_R, [30 to 720 min]. The equation was also tested and found satisfactory on a watershed in Illinois [23].

The U.S. Geologic Survey compared data from Wichita, Kansas against a relationship by Putnam, and found satisfactory results [30]. Putnam notes that "the estimates are most reliable for smaller size floods at sites where the drainage area ranges between 0.8 and 390 square kilometers [0.3 and 150 mi^2]

where the $L/\sqrt{S}$ ratio ranges between 11.6 and 10.50 [0.1 and 9.0] and where impervious cover of less than 30 percent is uniform and distributed over the Basin" [24]:

$$t_p = 0.0453 \left(\frac{L}{S^{0.5}}\right)^{0.5} I_w^{-0.57} \qquad \left[t_p = 0.49 \left(\frac{L}{S^{0.5}}\right)^{0.5} I_w^{-0.57}\right]$$

(6.32)

where L = length of the main watercourse km [mi]
       S = slope, m/m [ft/mi]
       I_w = impervious area/total area

Another method follows from the original Snyder equations, except that

$$C_t = \frac{7.81}{(I_a)^{0.78}} \qquad r_2 = 0.95 \text{ (coefficient of determination)} \qquad (6.33)$$

where I_a = percent of watershed which is impervious
         (for 90 percent impervious I_a = 90)

This equation was developed in 1975 as a revision to earlier information in the Denver *Urban Storm Drainage Criteria Manual* [25]. Figure 6.11 presents the equation. This relationship was developed by Colorado State University after studying new gaged rainfall-runoff relationships in the Denver area [29]. Also, many other points were developed based on data from other areas as indicated in Fig. 6.11. However, it is noted that there is significant scatter for areas less than 10 percent impervious; thus, it is not advisable to use this curve for such areas.

Also, the reader is reminded that a reasonableness check can be made by calculating 60 percent of the time of concentration as calculated by evaluation of flow time through the basin.

Estimation of the Synthetic Unit Hydrograph Peak q_p

The Snyder Equation for q_p is a function of t_p and a coefficient C_p; however, empirical correlations to various parameters are quite poor.

The most reasonable and recommended approach found was to use a basic relationship provided by the Tulsa District Army Corps as shown in Fig. 6.12. The data discussed earlier by Van Sickle [21] and Eagleson [22] were also plotted against this relationship and found to be in agreement. Note that the data here are also in British units. The equation of the relationship is:

$$q_p = 4.137 t_p^{-0.92} \qquad \left[q_p = 380 t_p^{-0.92}\right]$$

(6.34)

where q_p is in units of $m^3/(sec)(km^2)$ [$ft^3/(sec)(mi^2)$]. This relationship is similar to Eq. (6.18).

All of the data presented are found by determining the best fit of derived unit hydrographs (which are found by fitting effective rainfall and runoff flow data to the synthetic unit hydrograph equations). Thus, the good relationship shown is not surprising, and one can draw the conclusion that the accurate determination of the lag time discussed earlier is critical.

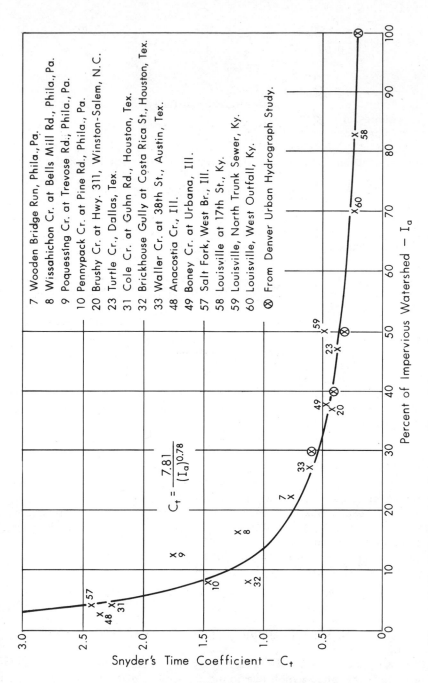

Figure 6.11 Relationship between C_t and imperviousness.

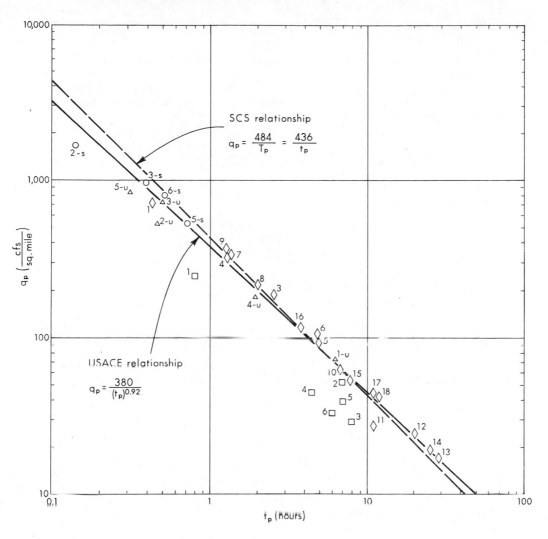

Figure 6.12 Synthetic Unit Hydrograph Procedure coefficients (t_p vs. q_p).
Note: For explanation of data points, see Fig. 6.10b.

Reference 21 poses equations for smaller basins of:
Rural:

$$q_p = (1.91 \times 10^3)A^{-0.12}T_R^{-0.30} \qquad \left[q_p = (1.70 \times 10^3)A^{-0.12} T_R^{-0.30} \right]$$

$$(6.35)$$

Urban:

$$q_p = (2.10 \times 10^4)A^{-0.09}T_R^{-0.94} \qquad \left[q_p = (1.93 \times 10^4)A^{-0.09} T_R^{-0.94} \right]$$

$$(6.36)$$

where T_R = time from the beginning of effective rainfall to the peak runoff
$\quad\quad$ A $\;$ = area in km^2 [mi^3]

The rural equation is based on watersheds with the following range: A, 0.35 to 18.2 km^2 [0.134 to 7.01 mi^2]; and T_R, 30 to 150 min. The urban equation is based upon A, 0.03 to 240 km^2 [0.0128 to 92 mi^2]; and T_R, 30 to 720 min. The equations have acceptable correlations, but not as good as the previously discussed equations for T_R.

Synthetic Unit Hydrograph Shape

The shape of the unit hydrograph is a function of the physical characteristics of the drainage basin. The shape is developed from empirical relationships such as those discussed previously and as follows. The following equations may be used for urbanized areas to estimate the width of the unit hydrograph at 50 percent and 75 percent of the peak discharge [25]. These values are similar to those proposed by Eagleson [22]:

$$W \text{ @ } 50\% \, Q_p = \frac{5.47}{q_p} \; (hr) \qquad \left[W \text{ @ } 50\% \, Q_p = \frac{500}{q_p} \right] \tag{6.37}$$

$$W \text{ @ } 75\% \, Q_p = \frac{2.84}{q_p} \; (hr) \qquad \left[W \text{ @ } 75\% \, Q_p = \frac{260}{Q_p} \right] \tag{6.38}$$

Army Corps numerator values (British units) for natural watersheds are 5.14 [470] and 3.06 [280], respectively, and indicate a more rounded peak [27].

Reference 21 provides guidance equations for small watersheds previously discussed:

	Rural		Urban	

$$W \text{ @ } 50\% \, Q_p = \frac{7.615}{q_p^{1.13} A^{0.02}} \quad \left[\frac{1230}{q_p^{1.13} A^{0.02}} \right] \quad\quad \frac{6.35}{q_p^{1.04} A^{0.01}} \quad \left[\frac{690}{q_p^{1.04} A^{0.01}} \right] \tag{6.39}$$

$$W \text{ @ } 75\% \, Q_p = \frac{4.805}{q_p^{1.13} A^{0.07}} \quad \left[\frac{740}{q_p^{1.13} A^{0.07}} \right] \quad\quad \frac{3.25}{q_p^{0.94} A^{0.02}} \quad \left[\frac{223}{q_p^{0.94} A^{0.02}} \right] \tag{6.40}$$

$$T_b \text{ (base)} = \frac{10.11 \, A^{0.11}}{q_p^{0.53}} \quad \left[\frac{123 \, A^{0.11}}{q_p^{0.53}} \right] \quad\quad \frac{34.94}{q_p^{1.19} A^{0.02}} \quad \left[\frac{7400}{q_p^{1.19} A^{0.02}} \right] \tag{6.41}$$

Once q_p is determined, Q_P (maximum unit hydrograph peak for the basin) can be computed by:

$$Q_p = q_p \, A \tag{6.42}$$

where A is the area of the basin in square miles.

The time from the beginning of rainfall to the peak of the unit hydrograph is determined by:

$$T_p = t_p + 0.5t_u \quad \text{(hr)} \tag{6.43}$$

Once Q_p is located, the unit hydrograph can be sketched with the aid of the approximate widths at $Q_{50\%}$ and $Q_{75\%}$. Sketching of the hydrograph can be assisted by comparison with the shape of the dimensionless unit hydrograph shown in Fig. 6.9. However, the area under the hydrograph should always be planimetered to determine the volume of runoff.

This volume should equal the volume of the unit runoff, i.e., 1-in. depth from the entire basin area A. If the two volumes are within 5 percent, then the sketched unit hydrograph is acceptable. The final step is to define the unit hydrograph in tabular form showing time vs. rate of flow. If Q_p does not fall on a chosen time interval so that the tabulation does not represent the graph, then the graph may be shifted so that the table will more truly represent the graph.

Design Storm Runoff

After the unit hydrograph has been calculated and the rainfall excess from the design storm determined, the design storm hydrograph can be calculated. The time units of the unit hydrograph abscissa should be the same as the time units of the rainfall excess, which for convenience should all be equal to the unit storm duration.

By multiplying the incremental rainfall amounts and the unit hydrograph values, a response can be obtained for each rainfall increment. Approximate lagging and addition of the responses from each rainfall increment result in the design storm hydrograph.

The SCS, in Ref. 12, presents the methodology for using the SCS synthetic triangular hydrograph as somewhat of a simplification to that discussed above.

SUHP Example

The SUHP example, presented in British units, determines the 100-year hydrograph for a developed basin of generally single-family homes (1/4-acre lots) in an area in central Oklahoma. The design rainfall excess has been previously determined and is presented in Table 6.10.

$$
\begin{aligned}
\text{Area} &= 2.0 \text{ mi}^2 \\
\text{L} &= 4.0 \text{ mi} \\
\text{L}_{ca} &= 2.5 \text{ mi} \\
\text{Pervious area} &= 60 \text{ percent} \\
\text{Impervious area} &= 40 \text{ percent} \\
\text{Slope} &= 1 \text{ percent} = 53 \text{ ft/mi}
\end{aligned}
$$

Step 1.   Determine t_c.

a. Using the Army Corps equation [Eq. (6.29)] for 100 percent urbanization:

$$t_p = 0.59 \left(\frac{LL_{ca}}{\sqrt{S}} \right)^{0.39} = 0.59 \left(\frac{4 \times 2.5}{\sqrt{53}} \right)^{0.39} = 0.67 \text{ hr}$$

Table 6.10 Determination of Storm Hydrograph

Time (min)	Unit hydrograph (ft^3/sec)	0.01	0.02	0.07	0.14	0.26	0.56	1.26
0	0							
10	150							
20	320							
30	700	0						
40	1130	2	0					
50	1160	3	3	0				
60	900	7	6	12	0			
70	680	11	14	26	22			
80	540	12	22	56	44	39	0	
90	420	9	23	90	98	83	84	0
100	330	7	18	93	158	182	179	189
110	280	5	13	72	162	294	392	403
120	220	4	11	54	126	302	633	882
130	170	3	8	43	94	234	650	1424
140	140	3	7	34	76	177	504	1460
150	100	2	6	26	58	140	381	1134
160	80	2	4	22	46	109	302	857
170	60	1	3	18	40	86	235	680
180	40	1	3	14	30	73	185	529
190	20	1	2	11	24	57	157	416
200	0	1	2	8	20	44	123	353
210		0	1	6	14	36	95	277
220		0	1	5	12	26	78	218
230			0	3	8	21	56	176
240			0	1	6	16	45	126
250				0	3	10	34	101
260					0	5	22	76
270						0	11	50
280							0	25
290								0
300								
310								
320								
330								
340								
350								
360								

0.55	0.28	0.19	0.17	0.15	0.09	0.08	0.07	Design storm hydrograph (ft^3/sec)
								0
								0
								0
								0
								2
								6
								25
								73
								173
								387
0								826
28	0							1360
176	42	0						2230
385	90	29	0					2960
622	196	61	26	0				3166
638	316	133	54	23	0			2911
495	325	215	119	48	14	0		2558
374	252	220	192	105	29	12	0	2247
297	190	171	197	170	63	26	11	1960
231	151	129	153	174	102	56	22	1686
182	118	103	116	135	104	90	49	1448
154	92	80	92	102	81	93	79	1202
121	78	63	71	81	61	72	81	968
94	62	53	56	63	49	54	00	759
77	48	42	48	50	38	43	47	587
55	39	32	37	42	30	34	38	455
44	28	27	29	33	25	26	29	344
33	22	19	24	26	20	22	23	250
22	17	15	17	21	15	18	20	170
11	11	11	14	15	13	14	15	104
0	6	8	10	12	9	11	12	68
	0	4	7	9	7	8	10	45
		0	3	6	5	6	7	27
			0	3	4	5	6	18
				0	2	3	4	9
					0	2	3	5
						0	1	1

b. Using Espey's equation [Eq. (6.31] for urban areas,

$$T_R = 20.8UL_f^{0.29} S_f^{-0.11} I^{-0.61}$$

$$= \frac{20.8(0.8)(4 \times 5280)^{0.29}}{(0.01)^{0.11}(40)^{0.61}} = 52.2 \text{ min} = 0.87 \text{ hr}$$

but rearranged from Eq. (6.27),

$$t_p = T_R - \frac{t_u}{2} = 0.87 - 0.08 = 0.75$$

(*Note*: see t_u in step 4, below.)

c. Using Eq. (6.32) by Putnam,

$$t_p = 0.49 \left(\frac{L}{\sqrt{S}} \right)^{0.5} I_w^{-0.57} = 0.49 \left(\frac{4}{\sqrt{53}} \right)^{0.5} 0.4^{-0.57} = 0.61 \text{ hr}$$

d. Using the suggested Eq. (6.33) for C_t,

$$C_t = \frac{7.81}{(I_a)^{0.78}} = \frac{7.81}{(40)^{0.78}} = 0.44$$

and then the Snyder equation [Eq. (6.15)],

$$t_p = C_t (LL_{ca})^{0.3} = 0.44(4 \times 2.5)^{0.3} = 0.80 \text{ hr}$$

e. Recommend a value for t_p.

The 0.67 value given by the Army Corps Eq. (6.29) seems reasonable since it fits within the scatter of data points on Fig. 6.1 and is based on Oklahoma watersheds.
The 0.75 value from the Epsey Eq. (6.31) is reasonable, but note that the 1 percent slope of the basin being analyzed fits on the upper range of the data from which the equation was derived.
The 0.61 value given by Putman's Eq. (6.32) is reasonable, but because of the 40 percent impervious area it is out of the data range from which the equation was derived.
The 0.80 value from Eq. (6.33) and Snyder's Eq. (6.15) is reasonable, but is based on a wide range of watersheds
The 0.67 and 0.75 values are judged to be the most reasonable and 0.67 is recommended as being conservative.

Step 2. Determine Q_P.

a. From Fig. 6.12, q_p is 600 ft^3/(sec)(mi^3)(in.).
b. $Q_P = q_p A = 600 \times 2 = 1200$ ft^3/(sec)(in.).
c. Equation (6.36) from Ref. 21 results in $q_p = 480$ ft^3/(sec)(mi^3) (in.); thus the result above is reasonable.

Step 3. Determine the width of the unit hydrograph at 50 percent and 75 percent of the Q_P, from Eqs. (6.37) and (6.38).

$$W @ 50\% Q_P = \frac{500}{Q_P/A} = \frac{500}{600} = 0.83 \text{ hr} \tag{6.37}$$

$$W @ 75\% Q_P = \frac{260}{Q_P/A} = \frac{260}{600} = 0.43 \text{ hr}$$

Interestingly, Ref. 21 Eqs. (6.39), (6.40), and (6.41) give values of 0.88 hr and 0.54 hr for W at 50 percent Q_p and W at 75 percent Q_p, respectively; and for T_b give 3.6 hr.

Step 4. Determine the unit time increment t_u from Eqs. (6.21) and (6.22).

$$t_u = \frac{t_c}{7.5} = \frac{t_p/0.6}{7.5} = \frac{0.67/0.6}{7.5} = 0.15 \text{ hr} = 8.9 \text{ min}$$

Use t_u = 10 min.

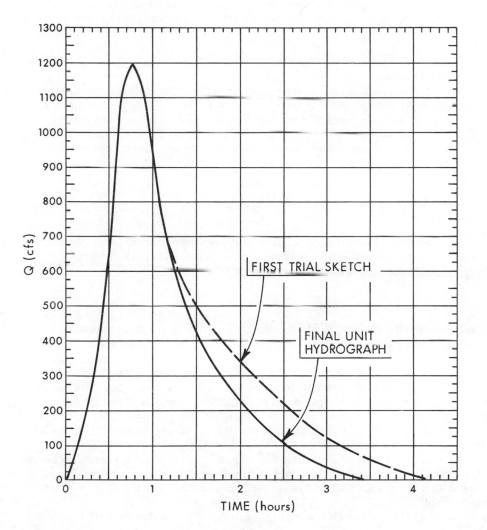

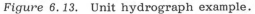

Figure 6.13. Unit hydrograph example.

Step 5. Determine the time-to-peak from the beginning of rainfall [Eq. (6.20)].

$$T_p = t_p + \frac{t_u}{2} = 0.67 \text{ hr} + \frac{0.167}{2} = 0.75 \text{ hr} \qquad \text{(rounded)}$$

Step 6. Using the results of steps 5, 6, and 7, sketch a unit hydrograph. See Fig. 6.13.

Step 7. The volume of the unit hydrograph should be 107 acre-ft, i.e., 1-in. runoff from 2 mi^2. A revision was made to lower the volume from the first estimate of 130 acre-ft to 109 acre-ft.

Step 8. Present the unit hydrograph in graphic form as shown in Fig. 6.13.

Step 9. Obtain the design excess precipitation values in unit duration increments. See the previous rainfall and rainfall excess example per SCS method. Remember the duration storm should be greater than $2 \times t_c = 2 \times t_p/0.06 = 2.24$ (i.e., use 3-hr storm).

Step 10. Set up Table 6.10.

Step 11. Multiply the precipitation value at the top of each column by each of the unit hydrograph ordinates and put the product in the corresponding time. Note that the first rainfall excess increment occurs during the period from 30 to 40 min; thus, the product hydrograph begins in time from 30 to 40 min. As the next increment of rainfall excess is 10 min later, the product hydrograph is also 10 min later.

6.5 MASSACHUSETTS INSTITUTE OF TECHNOLOGY CATCHMENT MODEL

One of the computer models mentioned in the previous chapter, the Massachusetts Institute of Technology Catchment Model (MITCAT) [31], is useful in evaluating drainage management programs.

Philosophy of the Catchment Model

MITCAT represents the physical movement of water over the catchment surface and through the channel network of a river basin [32]. Recognizing that surface geometry is extremely irregular and impossible (and uncessary) to represent in complete detail in either a physical or a mathematical model, MITCAT follows a reductionist approach. The model replaces the natural complexities with a number of simple elements such as overland flow planes, stream segments, pipe segments, etc. A suitable combination of an appropriate number of these simple elements is assumed sufficient to model the behavior of an entire catchment.

A sample catchment is illustrated in Fig. 6.14. A possible combination of overland flow and streamflow elements appears as a detailed model of this catchment in Fig. 6.15. Less detailed representations may be used in most practical applications.

Some of the basic considerations governing the development of the model were:

The model was based on sound physical reasoning.

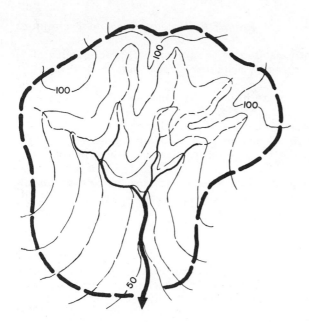

Figure 6.14. Drainage network of typical catchment. (Reproduced from Ref. 31.)

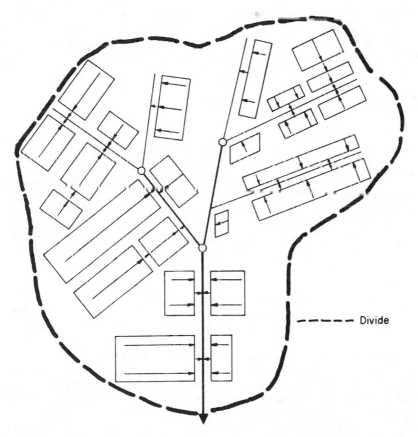

Figure 6.15. Equivalent "block" diagram of the catchment. (Reproduced from Ref. 31.)

The parameters were directly related to the physical characteristics of
the catchment and were directly measurable from map or field data
whenever possible.

"Fudge factors," which proliferate in many hydrologic models, were
avoided.

The model was made as simple as possible, and minimizes the amount of
field data required before reasonable results could be derived. The
results are, in general, not very sensitive to uncertainty in the
parameter estimates.

The model eliminates or minimizes the requirements for historical rainfall
or runoff data for calibration or estimation of parameters.

The model handles catchments of any size or shape, urban or rural, and
does not require complex input data to achieve this capability.

The model has been designed principally to simulate storm runoff events.

Numerical approximations, such as finite grid mesh sizes, optimum time
step, linearization procedures, etc., are controlled internally to make
the model as self-sufficient as possible and to make the model useful
to engineers not familiar with the numerical analysis or numerical
solution of different equations.

With these considerations in mind, certain basic runoff elements were
chosen. The elements considered were:

Flow distributed over the surface of the catchment is modeled by planes
of overland flow. An overland flow plane is subject to spatially uniform lateral inflow from rainfall, lateral outflow to infiltration, and
upstream inflow from adjacent overland flow segments.

Flow from the overland flow planes would be collected by streamflow
segments as lateral inflow and then passed downstream to other stream
segments. The term *stream* is used in a generic sense and presently
includes both open channel flow reaches and closed pipe elements.

Storage is simulated by reservoir elements.

Each of these elements can be linked together to simulate a prototype
basin. The elements are first linked into *segments* to define the model of a
single subbasin. These segments are in turn linked together to define the
total basin network. The choice and arrangement of elements in any particular situation depends upon several factors, which are presented in subsequent
sections.

Model Overview

The present MITCAT model can input several forms of rainfall data, which are
then reduced to effective rainfall by one of several infiltration algorithms.
Holtan's, Horton's, an Antecendent Precipitation Index Method, SCS, or direct
input methods are immediately available for use; other algorithms have been
programmed but are not directly on line.

Runoff as overland flow and as stream is simulated by use of the
kinematic wave equation. The kinematic wave model was adopted as the prime
routing method due to several factors. As the simplest of the nonlinear
models available, it yields the benefits of a nonlinear response without needing
an unduly complicated solution procedure. The use of any equation of hydraulic unsteady flow in modeling the phenomenon of overland flow is subject
to considerable parameter adjustment to account for the peculiar flow regime

involved. To this end, the two parameters of the kinematic model yield sufficient range to model the runoff phenomenon.

The kinematic wave equation for an overland flow segment is:

$$\frac{\partial y}{\partial t} + \frac{\partial q}{\partial x} = i - f \tag{6.44}$$

$$q = \alpha_c y^{m_c} \tag{6.45}$$

where y is the depth of flow, q is the rate of flow, t is time, x is distance along the segment, i is the rainfall intensity, and f is the infiltration rate. In Eq. (6.44), both i and f may vary with x and t. The difference i - f may be treated as an effective rainfall rate (which, by convention in hydrology, is never negative). The fact that f may vary with x causes the model to simulate runoff only from those locations where i exceeds f.

The corresponding equation for the stream segments is:

$$\frac{\partial A}{\partial t} + \frac{\partial Q}{\partial x} = q \tag{6.46}$$

$$Q = \alpha_s A^{m_s} \tag{6.47}$$

where A is the cross-sectional area of flow, Q is the discharge rate, and q is the lateral inflow rate of overland flow.

The above equations contain the so-called kinematic wave parameters (α, m) for both overland flow and stream segments. The estimation of these parameters is performed internally in the model from readily derived properties of the element. These properties generally include:

    Cross section shape
    Slope
    Roughness factor, generally Manning's n

The extraction of the appropriate values of α and m are presented in the principal reference [31].

The model solves the kinematic wave equations by numerical techniques. The details of these techniques have been carefully developed over a period of many years to the point where reliable procedures have been programmed to automatically assure the most economical solution of these equations [31].

The data input needs can generally be classified as follows:

    Rainfall
    Infiltration parameters
    Area of basin or subbasin
    Length of typical overland pattern
    Slope of typical overland runoff pattern
    Roughness of typical overland runoff pattern
    Impervious cover of basin
    Length of stream through basin
    Slope of stream through basin
    Roughness of stream through basin
    Typical section of stream through basin

The analysis and reduction of a basin to a representative model is a reasonable process and is the final determination as to the extent of data necessary to fulfill the above classes.

The full MITCAT model, as currently operational, requires approximately 350K bytes of core to run. A small nonproprietary version, referred to as MINICAT, is available which has basic capabilities of watershed representation, routing elements, and two infiltration methods (SCS or Horton's); MINICAT requires 140K bytes to operate, though this requirement can be reduced to 64K bytes.

MITCAT Structure and File System

The key elements of MITCAT are its modular structure and extensive file system. These allow flexibility for the user to decide how to model a basin and how to analyze the problem under study.

Structure of the Program

MITCAT follows certain basic steps during a simulation run. These are:

> *Read and check input data*: The user's commands and the associated data are read in and checked for errors, and if any are present, messages are printed out.
> *Display input data*: All of the input data, as well as data picked up from the file system or estimated by MITCAT, are printed out. This allows easy verification of the input data by the user.
> *Routing and output*: Each active element in the system is then simulated and its results are printed/plotted/stored.

MITCAT operates on the basin configuration in a sequential manner. This means that the whole time history of interest is processed through each of the simulating modules in sequence. This mode of operation should be contrasted with the parallel modes where the state (i.e., water stage, storage, etc., rather than discharges) of the total system is computed at each time step.

Use of a sequential model has the advantage that each individual element can be operated at different time steps. The size of the time increment may be dictated by model stability and/or convergence criteria or simply by the user's idea of what may be an "optimum" time step for such an element.

A disadvantage of this mode of operation is that decision making (where it is required) at each time step is rather difficult since the "state" of the system at all the downstream points is not known. On the other hand, a parallel model of operation permits ready decision making at each time step.

DATSYS File System

A feature of MITCAT is its associated file system (DATSYS). The data file handling system is designed to allow the user to access, process, etc. data which has been stored in the data bank system during the operation of a simulation (or other program). The outputs from these runs are usually time histories of data which are frequently necessary to plot (separately or as multiple plots), compute the moments of, and/or copy to other files. It is also possible to load historic data into the file system so that they may be used by a main line program for a simulation run.

Figure 6.16 shows the basic structure of the MITCAT System. Most important for the purpose of this book are the connections between MITCAT, the data bank, and the land use files. The efficient use of this file system can reduce the cost of a simulation run. For example, a run can be restarted at an intermediate point in a catchment if a new test must be made where only a change in the lower part of the basin is to be made.

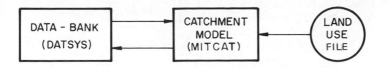

Figure 6.16. MITCAT system structure. (Reproduced from Ref. 31.)

Modeling Strategy

The general philosophy behind the catchment model was presented earlier in this section. As was emphasized, the model follows a reductionist approach using simple elements to model the natural complexities of a drainage basin. Presented here are the approach to be used in modeling the basin and the available methods of linking the various individual elements. Also presented are the strategies available to the user in modeling the behavior of complex hydrologic systems.

MITCAT is primarily a model to permit simulation of individual storms. It is thus more suitable for specific flood event simulation than for long-term simulation which might be required for low-flow determination or storage yield analysis. Another model, which may be based on the MITCAT data base, should be used if long-term simulation is desired.

MITCAT Elements

MITCAT is designed to operate as an interconnected series of simple routing elements. The operation of each of these elements is simulated by a relatively simple set of nonlinear techniques. These techniques were selected so that the data required for each could generally be derived from readily available sources such as topographic maps, soil surveys, and stream cross sections. At the same time, a determined effort was made to limit the number of black-box techniques and the resulting factors which would be required. The resulting MITCAT model is a deterministic simulation model with little calibration required to model the behavior of a given hydrologic system. Although the operation of each element within the model can be readily and easily simulated, the user must ensure that the chosen network of segments and elements does, in fact, represent the prototype hydrologic system in the basin. This section is designed to present some of the points considered germane to the correct choice of modeling elements and to present operational techniques developed in the first years of applied use of MITCAT.

Initial examination of the basin to be simulated will indicate the occurrence of a number of specific locations, both natural and those commonly referred to as "design points," which serve to define specific points of interest along the various stream reaches within the network. The natural points usually occur at locations such as stream junctions, highway crossings, and reservoir sites. The first step in the application of a limited-element model is to subdivide the basin into a number of subbasins. These are shown in Fig. 6.17, and are chosen so that subbasin boundaries occur at each of the following locations:

> Primary junctions
> Design points
> Stream flow gage locations (if available)
> Reservoir sites

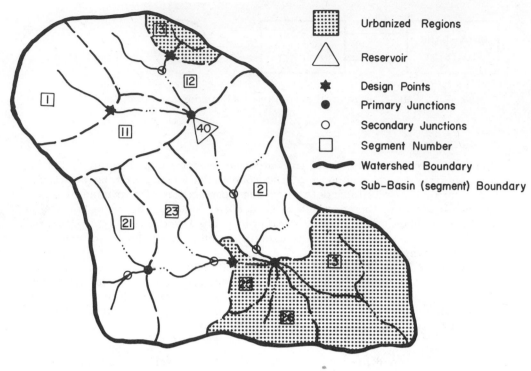

Figure 6.17. Watershed example. (Reproduced from Ref. 31.)

These initial subbasins may need to be further subdivided to account for major nonhomogeneities within them. For example, extreme rainfall variations due to orographic effects or development of part of a subbasin should encourage the modeler to subdivide further until a reasonably homogeneous condition exists. It is difficult to define what specific size limitations might be applied to the selection of the subbasins. Operational use of the model has indicated that subbasin sizes ranging up to 10 km^2 may be adequately modeled using a single model segment. A *segment* is the MITCAT model equivalent of a subbasin. In areas with relatively steep slopes and limited urbanization, segment sizes of up to 26 km^2 have been used. It is normally considered good practice, however, in situations where such large segments are involved to select the subbasins so that at least three subbasins are modeled upstream of any design points. This prevents the "lumping" inherent in the model application from significantly distorting the outflow hydrograph at the desired location.

The basic network of subbasins indicated in Fig. 6.17 can be visualized as in Fig. 6.18. This figure represents the MITCAT network connectivity. It is noted that such a schematic closely follows the prototype system with one notable exception. MITCAT programming considerations limit, to not more than two, the number of segments which may be linked as upstream inputs to a segment. In order to model the situation where segments 25, 26, and 2 are linked to segment 3, it would be necessary to insert a "dummy" segment, such as number 31, in the network. Such segments serve no routing purpose but are simply used to permit simulation of networks which have junctions with numerous streams tributary at the junction.

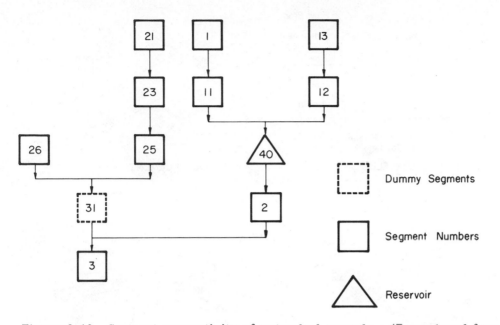

Figure 6.18. Segment connectivity of watershed example. (Reproduced from Ref. 31.)

The selection of the model *elements* required to simulate the hydrologic behavior within each subbasin is the next area where modeling strategy is involved. There are four such elements available within MITCAT, namely:

The *catchment* element, which is used to simulate overland flow
The *stream* element, which is used to simulate the behavior of a stream reach
The *reservoir* element, which is used to simulate the operation of on-stream regulation facilities, and smaller storage systems such as reservoirs for site detention control
The *pipe* element, which may be used to model the behavior of closed conduits

Typical schematics of each of these elements are presented in Fig. 6.19.

Most subbasins (segments) are modeled using a simple combination of catchment and stream elements as illustrated in Fig. 6.20. The typical V-sloped segment shown has been found to be an extremely efficient and accurate way of modeling the behavior of far more complex prototype systems.

Catchment Element
In the V-sloped segment rainfall input is applied to each of the catchment elements. The operation of each of these elements is simulated by modeling the processes of infiltration and overland flow which occur on a unit strip of each element. In order to apply simple modeling techniques, all the parameters for each of these elements are considered to be spatially homogeneous. The parameter that is of greatest interest to the model at this stage is the overland flow length, designated by Lc_1 or Lc_2 in Fig. 6.20. Inspection of the topographic map of each subbasin will reveal some variations of this value

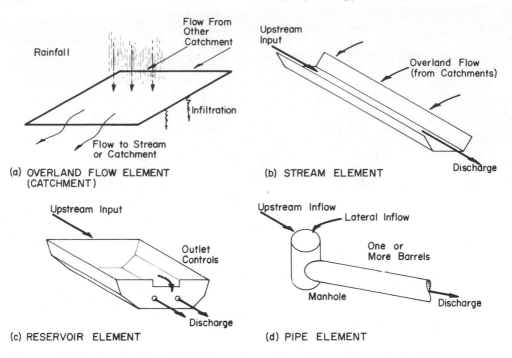

Figure 6.19. Typical MITCAT elements. (Reproduced from Ref. 31.)

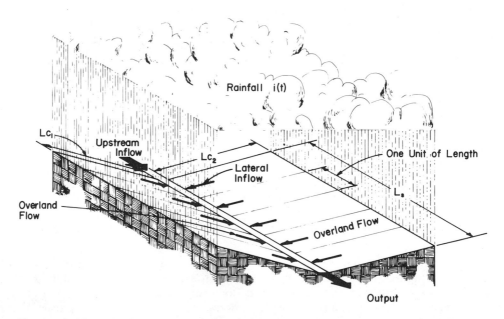

Figure 6.20. Simple catchment-stream elements used to model a segment. (Reproduced from Ref. 31.)

over the subbasin. It will be found, however, that for most natural basins, a surprisingly constant value of Lc can be found when one has determined what the full drainage network within the basin looks like. This network includes not only the main stem streams, but also the multitude of smaller tributaries and rivulets which act as channels under significant rainfall events. Such a schematic is shown in Fig. 6.21. The correct selection of the value(s) of this overland flow length Lc is essential for the adequate modeling of the basin response. This parameter is by far the most important variable for determining the response characteristic of the overland flow segment. Since these elements in turn act as the primary high frequency and limited bandwidth filters in the simulation network, their accurate representation is crucial to the ability of the chosen schematic to simulate the prototype system.

Table 6.11 gives Manning's n for overland flow roughness effects. Since the selection of the correct overland flow length is so critical, various internal arrangements of the segment structure will be required in order to adequately include the range of overland flow lengths which may occur in the prototype basin. In certain cases, a very simple schematic as shown in Fig. 6.22a will suffice. The more general case is shown in Fig. 6.22b, where an overland flow element is modeled in each side of the main stem. In this case, the overland flow lengths for the elements may be the same or, as more generally found, somewhat different. This schematic is also able to handle the normal variations of the other parameters such as slope and roughness as they occur.

A more complex configuration is illustrated in Fig. 6.22c. This would be used to model an urban segment where a number of different flow situations exist (Fig. 6.23). Elements would include:

Flow from impervious roadway and roof surfaces
Flow from a pervious lawn accepting runoff from commercial areas
Flow from impervious commercial/parking areas

Roof drains normally discharge directly into pipe systems in Melbourne, and so roof and street areas are combined into one element. Figure 6.23c is

Table 6.11 Manning's n for Overland Flow

Surface	n
Dense growth	0.4 -0.5
Pasture	0.3 -0.4
Lawns	0.2 -0.3
Concrete/asphalt:	
Shallow depths	0.10-0.15
Small depths	0.05-0.10[a]

[a]Significant flow occurring over the surface.
Source: U.S. Forest Service [13].

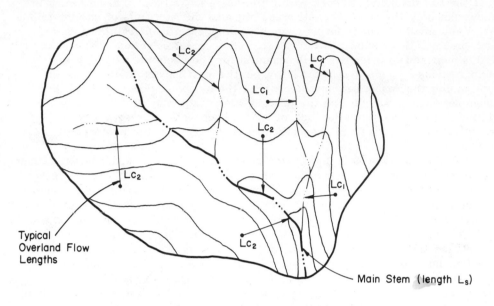

(a)   PROTOTYPE WATERSHED

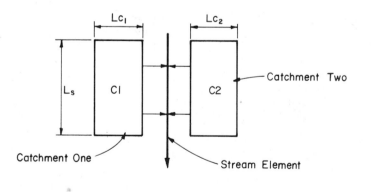

(b)   MODEL SCHEMATIC

Figure 6.21.   (a)   Prototype watershed; (b) typical model schematic for natural watershed.   (Reproduced from Ref. 31.)

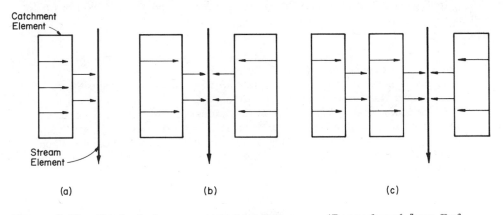

Figure 6.22. Typical stream segment structures. (Reproduced from Ref. 31.)

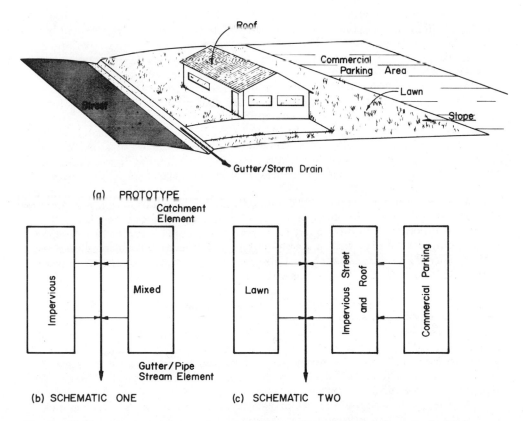

Figure 6.23. Typical model of urban area. (Reproduced from Ref. 31.)

presented to illustrate a more complex model for some applications. Segments
shown in Fig. 6.23b and c have been successfully used for subbasins up to
400 hectares (1000 acres) in size.

Stream Element

In each of the schematics, a *stream* element is considered to be a stream
reach of uniform cross section and to have a length of Ls.

The cross section is usually considered to be a simple shape as shown in
Fig. 6.24 where the basic shapes currently available for simulation with
MITCAT are illustrated. The model is not especially sensitive to the cross
section chosen where discharge simulation is involved; however, stage,
velocity, and other derived data are sensitive to the shape chosen. From
operational use of MITCAT, some general rules have emerged for the appropriate section shapes to use.

It is found that most upstream reaches are best simulated by use of
simple triangular shapes. Many of these channels are extremely small, and
any flood flow uses the overbank area extensively. Such overbank areas are
frequently triangular in cross section—see Fig. 6.25a.

In further downstream reaches, simulation with basically trapezoidal
channel sections is found to be most appropriate—see Fig. 6.25b. In such
cases, care must be taken to fit the most appropriate cross-section shape for
the range of flood stages of interest.

Urban areas sometimes have gutter sections as their primary drain
element. In these segments, use of a gutter shape as in Fig. 6.25c will be
desirable. However, for the simulation of larger urban areas, where storm
drains are used, a stream segment with a circular shape (Fig. 6.24e) should
be used. This particular shape actually simulates the behavior of a regular
conduit up to the 90 percent full position. Flows greater than that are
modeled by assuming that the capacity of the element increases as required,
i.e., no pressure flow and/or surcharging occurs. Use of a circular cross
section is appropriate since the area being modeled does not normally drain
to a single storm drain; rather, it consists of a whole series of storm drains
connected in a relatively complex network. However, it has been found
possible to duplicate the whole network by a *single* stream/pipe element as
shown in Fig. 6.23b or c.

As an area is urbanized, its drainage network simulation will also change.
A typical staged development of a segment schematic is shown in Fig. 6.26.
In this situation, the undeveloped (natural) basin can be typically modeled
using a single segment, with a simple catchment-stream-catchment structure
of the segment. As the basin is urbanized, more and more of the area is provided with storm drains (including gutters) and a different set of drainage
controls exists. The easiest way to simulate the runoff from such urbanizing
areas is by using a segment in parallel to the segment modeling the runoff
from the natural system—see Fig. 6.26b. The discharges from both of these
segments are combined to estimate the total flow from the basin. In the fully
developed case, the runoff from the total basin may well be simulated by the
urban segment and only the upstream flow routed through the main stem.
The latter may, of course, also be modified in order to handle the increased
discharge levels, and so forth.

As discussed above, a range of relatively simple cross-section shapes
are used to model prototype stream channels. These shapes are normally
sufficient to enable adequate determination of the discharge rate, and were

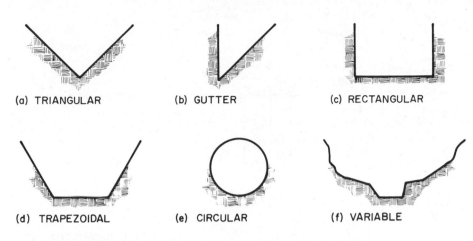

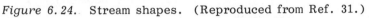

Figure 6.24. Stream shapes. (Reproduced from Ref. 31.)

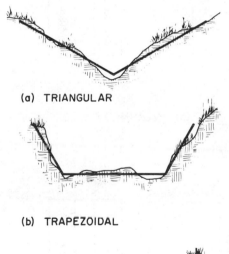

(a) TRIANGULAR

(b) TRAPEZOIDAL

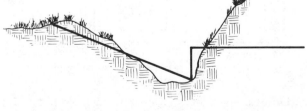

(c) GUTTER

Figure 6.25. Selection of stream cross section. (Reproduced from Ref. 31.)

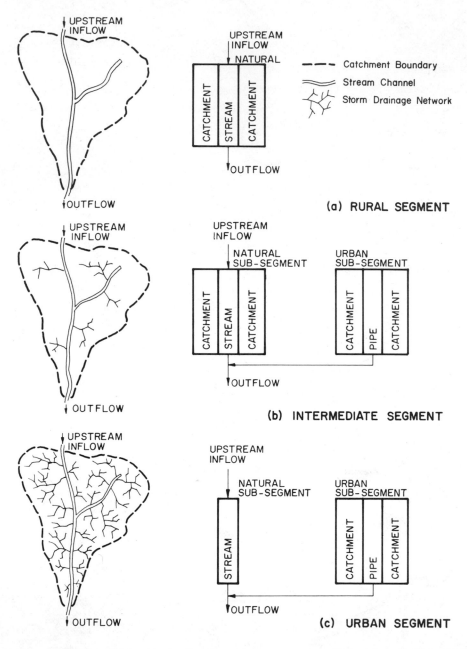

Figure 6.26. Typical urbanization of a segment. (Reproduced from Ref. 31.)

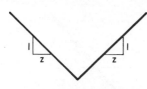

TRIANGULAR

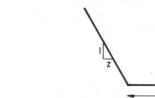

GUTTER

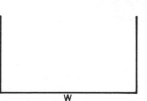

RECTANGULAR

TRAPEZOIDAL

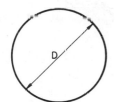

CIRCULAR

VARIABLE

W : Bottom Width
Z : Side Slope (z horizontal to I vertical)
D : Diameter
● : Each Point on Variable Section Defined
    by Station and Elevation Coordinates

Figure 6.27. Stream cross-section definitions. (Reproduced from Ref. 31.)

chosen because simulation with such simple elements is extremely fast. They will not, of course, be sufficiently detailed to allow accurate determination of the equivalent flow stages. In this instance, other techniques are available within MITCAT to permit adequate stage determination.

Variable. Available cross sections are illustrated in Fig. 6.27 where their significant parameters are noted. It should be noted that the parameters used include the cross-section shapes as well as the segment slope S and the effective Manning n. The value of n appropriate for stream segments can be estimated from field inspection or from standard hydraulic handbooks.

Triangular Sections. These sections will be very useful for simulating *small* natural streams, i.e., where permanent flow is very small compared to flood flows, and streams that flow through unimproved valleys. Certain man-made channels may also fit this category.

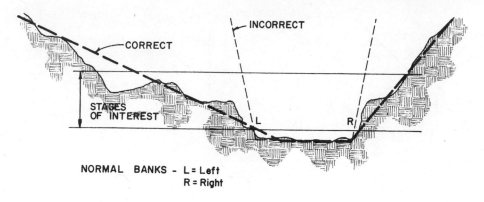

NORMAL BANKS - L = Left
                        R = Right

(a) SIGNIFICANT OUT-OF-BANKS FLOW

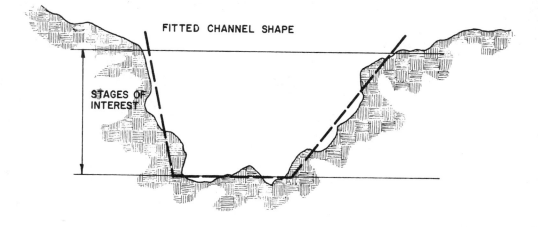

(b) CONSTRICTED SECTIONS

Figure 6.28. Fitting trapezoidal cross section to natural channels. (Reproduced from Ref. 31.)

Gutter Sections. These sections are of use in urban areas where significant flow in the gutters is to be modeled.

Rectangular Section. This section shape is of most use in simulating man-made stream segments where the flow is expected to stay in banks for the event being considered. It is preferable to use either a triangular or a trapezoidal section to model most prototype conditions.

Trapezoidal Sections. This section shape is of interest in modeling the stream channels occurring in the downstream part of river basins. It is very important in this instance to choose the correct side slopes to adequately handle the full range of flow stages. Figure 6.28 illustrates the choices available.

Circular Sections. Modeling the storm drainage system in urban areas requires the capability of simulating the response characteristics of stream segments with a circular cross-section slope. In many cases, the pipe will never fill completely during a storm event and the routing through the pipe will resemble normal stream routing. The method allows the pipe (or circular cross-section stream segment) to continue to respond even under surcharge conditions; i.e., there is no upper limit to the capacity of the stream segment. This method has proved very useful in simulating the response of urban basins without the need to specify the complete internal drainage system. Instead, a "typical" segment is specified and allowed to handle as much flow as required. To some extent, this duplicates the behavior of the prototype where runoff that is not accepted by catch basins usually flows over the surface until it either enters the storm drain system or is discharged to a natural watercourse.

Reservoir and Pipe Elements
The other simulating elements considered are the *reservoir* and *pipe* elements for which simple schematics are shown in Fig. 6.29a and b. Both elements allow modeling of on-line storage. The storage is considered explicitly in the reservoir elements. In pipe elements, each element (which can consist of a number of individual barrels) has an upper limit on its discharge

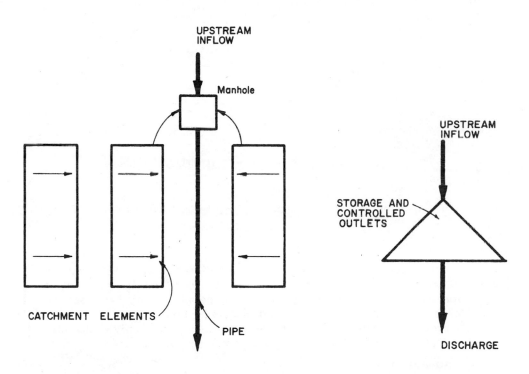

(a) PIPE SEGMENT WITH PIPE AND CATCHMENTS   **(b) RESERVOIR SEGMENT**

Figure 6.29. Typical reservoir and pipe segments. (Reproduced from Ref. 31.)

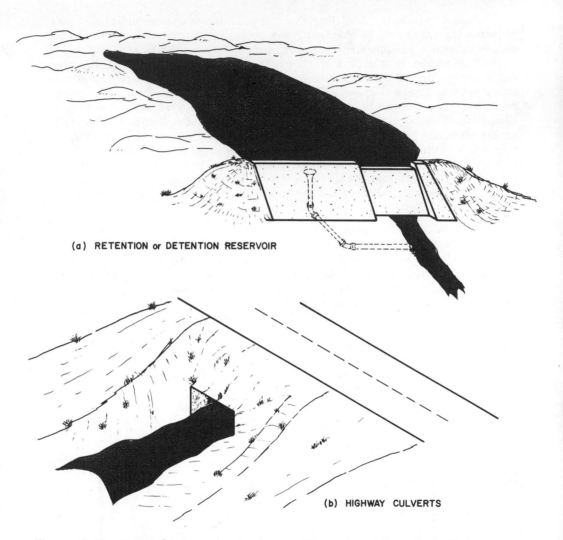

(a) RETENTION or DETENTION RESERVOIR

(b) HIGHWAY CULVERTS

Figure 6.30. Uses of reservoir elements. (Reproduced from Ref. 31.)

capacity. The model thus allocates a "manhole" storage volume upstream of the element to regulate flow into the pipe element. All flows in excess of the pipe's capacity are stored and released to the pipe as capacity becomes available. The pipe element will be discussed in detail later.

Reservoir elements are used to model a number of commonly occurring situations in the basin—see Fig. 6.30. These normally include:

Retention or detention reservoir
Highway culverts, with the embankment acting to form a storage volume under flood conditions

The operation of each of these systems can normally be simulated through the use of relatively simple rules.

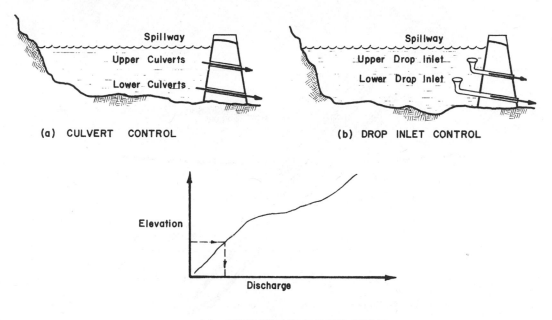

(a) CULVERT CONTROL

(b) DROP INLET CONTROL

(c) DISCHARGE/ELEVATION CURVE

Figure 6.31. Reservoir outlet controls. (Reproduced from Ref. 31.)

The outlet control conditions which can currently be simulated are shown in Fig. 6.31. Basically, the controls can consist of up to two outlets, each of which is either a simple pipe or a drop-inlet (glory hole) type. Combined with these is an emergency spillway. The operation of these outlets is solely a function of the water surface elevation in the reservoir at any given time. In order to provide the ability to simulate more complex conditions, the user can specify a discharge vs. elevation relation which the model is to follow. The above set of controls has been found adequate to simulate the operation of all the major on-stream reservoirs found in urban drainage systems. Routing through such reservoirs is based on a normal input/output/storage relationship, i.e., the discharge at a given time is solely a function of the available "head" in the reservoir and the characteristics of the outlet controls.

A pipe element could be useful for the detailed modeling of small urbanized areas—see Fig. 6.32. In these situations, the user may need to know the flow and grade line elevation at each manhole within the system—the pipe element provides this ability. The pipe element can consist of a number of parallel barrels of pipe section, each with a similar cross section.

The schematic shown in Fig. 6.32b shows that a pipe element can have lateral inflow from adjacent catchment elements. This lateral flow is not uniformly distributed along the pipe's length, as would be the case with a stream segment, but the total flow is input to the manhole which is assumed to be at the upstream end of each pipe. The manhole itself has a number of parameters (such as area and elevation) associated with it so that the model can adequately simulate the processes such as surcharging and spill to adjacent areas.

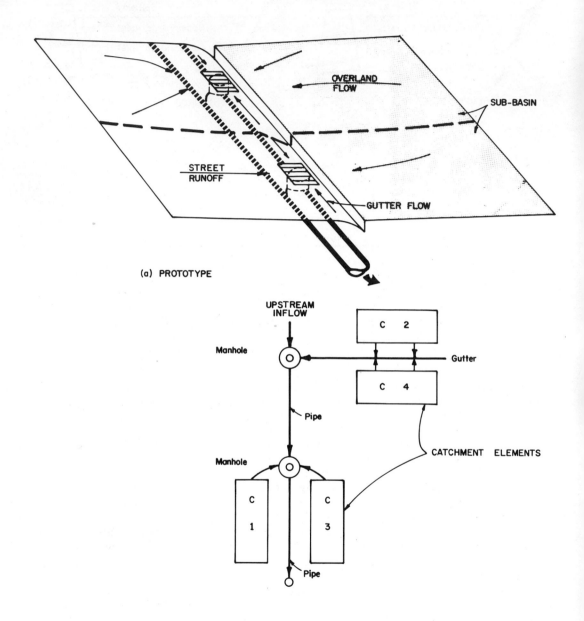

(a) PROTOTYPE

(b) MODEL OF TWO TYPICAL SCHEMATICS

Figure 6.32. Use of pipe elements. (Reproduced from Ref. 31.)

Other Capabilities

MITCAT has other algorithms which are useful in various situations. These include situations such as:

Off-stream storage
Splitting flow
Backwater situations
Many small streams entering a major stream
Stage discharge checks

These and other unusual situations are covered in the user's manual [31].

6.6 RUNOFF HYDROLOGY BY STATISTICAL ANALYSIS

Introduction

Where drainage basins are largely unchanged and have streamflow records, flood frequencies can be assigned by statistical methods. The following log-Pearson Type III Method exemplifies one such technique.

Log-Pearson Type III

The Pearson Type III Method was originally presented by H. A. Foster in 1924 [33]. As used by Foster, the method required the use of the natural data in computations of the mean, standard deviation, and skew coefficient of the distribution. The current practice is first to transform the natural data to their logarithms and then to compute the statistical parameters. Because of this transformation, the method is now called the log-Pearson Type III Method.

The following symbols are used in the log-Pearson Type III Method:

Y = arithmetic magnitude of an annual flood event
X = logarithmic magnitude of Y
N = number of events in the record being used
M = mean of the X values
$x = X - M$
S = standard deviation of the X values
g = skew coefficient
K = Pearson Type III coordinates expressed in number of standard deviations from the mean for various recurrence intervals or percent chance
Q = computed flood flow for a selected recurrence interval or percent chance

The events considered here are flood flows in the *annual series*. The physical units used for Y (such as cubic feet per second) are also those for Q.
The outline of work is as follows:

1. Transform the list of N annual flood magnitudes Y_1, Y_2, ..., Y_N to a list of corresponding logarithmic magnitudes X_1, X_2, ..., X_N.
2. Compute the mean of the logarithms:

$$M = \frac{\Sigma X}{N}$$

Table 6.12 K Values for Positive Skew Coefficients (Recurrence Interval in Years)

Skew coefficient (g)	1.0101	1.0526	1.1111	1.2500	2	5	10	25	50	100	200
	Percent chance										
	99	95	90	80	50	20	10	4	2	1	0.5
3.0	-0.667	-0.665	-0.660	-0.636	-0.396	0.420	1.180	2.278	3.152	4.051	4.970
2.9	-0.690	-0.688	-0.681	-0.651	-0.390	0.440	1.195	2.277	3.134	4.013	4.909
2.8	-0.714	-0.711	-0.702	-0.666	-0.384	0.460	1.210	2.275	3.114	3.973	4.847
2.7	-0.740	-0.736	-0.724	-0.681	-0.376	0.479	1.224	2.272	3.093	3.932	4.783
2.6	-0.769	-0.762	-0.747	-0.696	-0.368	0.499	1.238	2.267	3.071	3.889	4.718
2.5	-0.799	-0.790	-0.771	-0.711	-0.360	0.518	1.250	2.262	3.048	3.845	4.652
2.4	-0.832	-0.819	-0.795	-0.725	-0.351	0.537	1.262	2.256	3.023	3.800	4.584
2.3	-0.867	-0.850	-0.819	-0.739	-0.341	0.555	1.274	2.248	2.997	3.753	4.515
2.2	-0.905	-0.882	-0.844	-0.752	-0.330	0.574	1.284	2.240	2.970	3.705	4.444
2.1	-0.946	-0.914	-0.869	-0.765	-0.319	0.592	1.294	2.230	2.942	3.656	4.372
2.0	-0.990	-0.949	-0.895	-0.777	-0.307	0.609	1.302	2.219	2.912	3.605	4.298
1.9	-1.037	-0.984	-0.920	-0.788	-0.294	0.627	1.310	2.207	2.881	3.553	4.223
1.8	-1.087	-1.020	-0.945	-0.799	-0.282	0.643	1.318	2.193	2.848	3.499	4.147

1.7	4.069	3.444	2.815	2.179	1.324	0.660	-0.268	-0.808	-0.970	-1.056	-1.140
1.6	3.990	3.388	2.780	2.163	1.329	0.675	-0.254	-0.817	-0.994	-1.093	-1.197
1.5	3.910	3.330	2.743	2.146	1.333	0.690	-0.240	-0.825	-1.018	-1.131	-1.256
1.4	3.828	3.271	2.706	2.128	1.337	0.705	-0.225	-0.832	-1.041	-1.168	-1.318
1.3	3.745	3.211	2.666	2.108	1.339	0.719	-0.210	-0.838	-1.064	-1.206	-1.383
1.2	3.661	3.149	2.626	2.087	1.340	0.732	-0.195	-0.844	-1.086	-1.243	-1.449
1.1	3.575	3.087	2.585	2.066	1.341	0.745	-0.180	-0.848	-1.107	-1.280	-1.518
1.0	3.489	3.022	2.542	2.043	1.340	0.758	-0.164	-0.852	-1.128	-1.317	-1.588
0.9	3.401	2.957	2.498	2.018	1.339	0.769	-0.148	-0.854	-1.147	-1.353	-1.660
0.8	3.312	2.891	2.453	1.993	1.336	0.780	-0.132	-0.856	-1.166	-1.388	-1.733
0.7	3.223	2.824	2.407	1.967	1.333	0.790	-0.116	-0.857	-1.183	-1.423	-1.806
0.6	3.132	2.755	2.359	1.939	1.328	0.800	-0.099	-0.857	-1.200	-1.458	-1.880
0.5	3.041	2.686	2.311	1.910	1.323	0.808	-0.083	-0.856	-1.216	-1.491	-1.955
0.4	2.949	2.615	2.261	1.880	1.317	0.816	-0.066	-0.855	-1.231	-1.524	-2.029
0.3	2.856	2.544	2.211	1.849	1.309	0.824	-0.050	-0.853	-1.245	-1.555	-2.104
0.2	2.763	2.472	2.159	1.818	1.301	0.830	-0.033	-0.850	-1.258	-1.586	-2.178
0.1	2.670	2.400	2.107	1.785	1.292	0.836	-0.017	-0.846	-1.270	-1.616	-2.252
0	2.576	2.326	2.054	1.751	1.282	0.842	0	-0.842	-1.282	-1.645	-2.326

Source: Data from Ref. 34.

Table 6.13 K Values for Negative Skew Coefficients (Recurrence Interval in Years)

Skew coefficient (g)	1.0101	1.0526	1.1111	1.2500	2	5	10	25	50	100	200
	— Percent chance										
	99	95	90	80	50	20	10	4	2	1	0.5
0	-2.326	-1.645	-1.282	-0.842	0	0.842	1.282	1.751	2.054	2.326	2.576
-0.1	-2.400	-1.673	-1.292	-0.836	0.017	0.846	1.270	1.716	2.000	2.252	2.482
-0.2	-2.472	-1.700	-1.301	-0.830	0.033	0.850	1.258	1.680	1.945	2.178	2.388
-0.3	-2.544	-1.726	-1.309	-0.824	0.050	0.853	1.245	1.643	1.890	2.104	2.194
-0.4	-2.615	-1.750	-1.317	-0.816	0.066	0.855	1.231	1.606	1.834	2.029	2.201
-0.5	-2.686	-1.774	-1.323	-0.808	0.083	0.856	1.216	1.567	1.777	1.955	2.108
-0.6	-2.755	-1.797	-1.328	-0.800	0.099	0.857	1.200	1.528	1.720	1.880	2.016
-0.7	-2.824	-1.819	-1.333	-0.790	0.116	0.857	1.183	1.488	1.663	1.806	1.926
-0.8	-2.891	-1.839	-1.336	-0.780	0.132	0.856	1.166	1.448	1.606	1.733	1.837
-0.9	-2.957	-1.858	-1.339	-0.769	0.148	0.854	1.147	1.407	1.549	1.660	1.749
-1.0	-3.022	-1.877	-1.340	-0.758	0.164	0.852	1.128	1.366	1.492	1.588	1.664
-1.1	-3.087	-1.894	-1.341	-0.745	0.180	0.848	1.107	1.324	1.435	1.518	1.581

-1.2	-3.149	-1.910	-1.340	-0.732	0.195	0.844	1.086	1.282	1.379	1.449	1.501
-1.3	-3.211	-1.925	-1.339	-0.719	0.210	0.838	1.064	1.240	1.324	1.383	1.424
-1.4	-3.271	-1.938	-1.337	-0.705	0.225	0.832	1.041	1.198	1.270	1.318	1.351
-1.5	-3.330	-1.951	-1.333	-0.690	0.240	0.825	1.018	1.157	1.217	1.256	1.282
-1.6	-3.388	-1.962	-1.329	-0.675	0.254	0.817	0.994	1.116	1.166	1.197	1.216
-1.7	-3.444	-1.972	-1.324	-0.660	0.268	0.808	0.970	1.075	1.116	1.140	1.155
-1.8	-3.499	-1.981	-1.318	-0.643	0.282	0.799	0.945	1.035	1.069	1.087	1.097
-1.9	-3.553	-1.989	-1.310	-0.627	0.294	0.788	0.920	0.996	1.023	1.037	1.044
-2.0	-3.605	-1.996	-1.302	-0.609	0.307	0.777	0.895	0.959	0.980	0.990	0.995
-2.1	-3.656	-2.001	-1.294	-0.592	0.319	0.765	0.869	0.923	0.939	0.946	0.949
-2.2	-3.705	-2.006	-1.284	-0.574	0.330	0.752	0.844	0.888	0.900	0.905	0.907
-2.3	-3.753	-2.009	-1.274	-0.555	0.341	0.739	0.819	0.855	0.864	0.867	0.869
-2.4	-3.800	-2.011	-1.262	-0.537	0.351	0.725	0.795	0.823	0.830	0.832	0.833
-2.5	-3.845	-2.012	-1.250	-0.518	0.360	0.711	0.771	0.793	0.798	0.799	0.800
-2.6	-3.889	-2.013	-1.238	-0.499	0.368	0.696	0.747	0.764	0.768	0.769	0.769
-2.7	-3.932	-2.012	-1.224	-0.479	0.376	0.681	0.724	0.738	0.740	0.740	0.741
-2.8	-3.973	-2.010	-1.210	-0.460	0.384	0.666	0.702	0.712	0.714	0.714	0.714
-2.9	-4.013	-2.007	-1.195	-0.440	0.390	0.651	0.681	0.683	0.689	0.690	0.690
-3.0	-4.051	-2.003	-1.180	-0.420	0.396	0.636	0.660	0.666	0.666	0.667	0.667

Source: Data from Ref. 34.

3. Compute the standard deviation of the logarithms:

$$S = \sqrt{\frac{\Sigma\, x^2}{N - 1}} = \sqrt{\frac{\Sigma\, X^2 - (\Sigma\, X)^2/N}{N - 1}}$$

4. Compute the coefficient of skewness:

$$g = \frac{N\, \Sigma\, x^3}{(N - 1)\,(N - 2)S^3} = \frac{N^2\, \Sigma\, X^3 - 3N\, \Sigma\, X\, \Sigma\, X^2 + 2\,(\Sigma\, X)^3}{N(N - 1)\,(N - 2)S^3}$$

5. Compute the logarithms of discharges at selected recurrence intervals or percent chance:

$$\log Q = M + KS$$

Take K from Table 6.12 or Table 6.13 for the computed value of g and the selected recurrence interval or percent chance. Log Q is the logarithm of a flood discharge having the same recurrence interval or percent chance.

6. Find the antilog of log Q to get the flood discharge Q.

REFERENCES

1. *Hydraulics of Runoff from Developed Surfaces*, Carl F. Izzard, Proceedings of the 26th Annual Meeting of the Highway Research Board, Washington, D.C., December 5-8, 1946.

2. *Storm Drainage Criteria and Design Manual*, City of Fort Worth Public Works Department, Knowlton-Ratliff-English Consulting Engineers, Fort Worth, Tex., 1967.

3. *Design and Construction of Sanitary and Storm Sewers*, ASCE Manual of Engineering Practice, No. 37, American Society of Civil Engineers, New York, 1969.

4. *Determination of Runoff for Urban Storm Water Drainage System Design*, Kurt W. Bauer, *Technical Record*, Vol. 2, Nos. 4 and 5, Southeastern Wisconsin Regional Planning Commission, April-May 1965.

5. *Airport Drainage*, Federal Aviation Agency, Washington, D.C., 1965.

6. *Practices in Detention of Urban Stormwater Runoff*, American Public Works Association, Special Report No. 43, 1974.

7. *Surface Runoff Phenomena*, *Part I, Analysis of the Hydrograph*, R. E. Horton, Horton Hydrological Laboratory, Publ. 101, Ann Arbor, Mich., 1935.

8. *Model of Watershed Hydrology*, USDA HL-70, Technical Bulletin No. 1435, U.S. Department of Agriculture, Agricultural Research Service, Washington, D.C., 1971.

9. "Hydrology, Part I, Watershed Planning," *National Engineering Handbook*, Section 4, *Hydrology*, U.S. Department of Agriculture, Soil Conservation Service, Washington, D.C., August 1972.

10. *Urban Hydrology for Small Watersheds*, Technical Release No. 55, U.S. Department of Agriculture, Soil Conservation Service, Washington, D.C., January 1975.

11. *Forest and Range Hydrology Handbook*, U.S. Forest Service, Washington, D.C., 1959.

12. *Volume and Rate of Runoff in Small Watersheds*, SCS-TP-149, U.S. Department of Agriculture, Soil Conservation Service, Washington, D.C., January 1968.

13. *Handbook on Methods of Hydrologic Analysis*, U.S. Forest Service, Washington, D.C., 1959.

14. *The Gain, Transfer, and Loss of Soil Water*, J. R. Philip, Water Resources Use and Management, Melbourne University Press, Melbourne, 1963.

15. "The Theory of Infiltration, 4, Sorptivity and Algebraic Infiltration Equations," J. R. Philip, *Soil Science*, Vol. 84, 1957, pp. 257-264.

16. "Modeling Infiltration during a Steady Rain," R. G. Mein and C. L. Larson, *Water Resources Research*, Vol. 9, 1973, pp. 384-394.

17. *Handbook of Applied Hydrology*, Ven Te Chow, McGraw-Hill, New York, 1964.

18. *Synthetic Unit-Graphs*, F. F. Snyder, Transactions of the American Geophysical Union, Vol. 19, Washington, D.C., 1938.

19. *Snyder's Synthetic Unit Hydrograph Coefficients, Central and Northeastern Oklahoma*, 1977, personal communication with Dale Reynolds and Weldon M. Gamel, U.S. Army Corps of Engineers, Tulsa District, Okla., March 1979.

20. *Hydrology for Engineers*, Ray K. Linsley, Jr., Max A. Kohler, and Joseph L. H. Paulhus, McGraw-Hill, New York, 1958.

21. *Study of Some Effects of Urbanization on Storm Runoff from a Small Watershed*, W. H. Espey, Jr., William Howard, Jr., Carl W. Morgan, and Frank D. Masch, Texas Water Development Board, Report 23, Austin Tex., August 1966.

22. *Unit Hydrograph Characteristics of Sewered Areas*, Peter S. Eagleson, Proceedings of the ASCE, HY2, American Society of Civil Engineers, New York, March 1962.

23. *Some Effects of Urbanization on Floods*, John B. Stall, Michael L. Terstriep, and Floyd A. Huff, ASCE Meeting Preprint 1130, New York, January 1970.

24. *Effect of Urban Development on Floods in the Piedmont Province of North Carolina*, Arthur L. Putnam, U.S. Geologic Survey Open File Report, Washington, D.C., 1972.

25. *Urban Storm Drainage Criteria Manual*, Volumes I and II, Denver Regional Council of Governments, Wright-McLaughlin Engineers, Denver, 1969; revised May 1975.

26. *The Application of Synthetic Unit Hydrographs to Drainage Basins in the Riverside County Flood Control and Water Conservation District*, Riverside, Calif., January 1963.

27. *Unit Hydrograph*, Civil Works Investigations, Project 152, U.S. Army Corps of Engineers, Baltimore District, Md., 1963.

28. *HEC-1, Flood Hydrograph Package, Users Manual*, U.S. Army Corps of Engineers, Hydrologic Engineering Center, Davis, Calif., January, 1973.

29. *Determination of Urban Watershed Response Time*, Hydrology Paper 71, E. F. Schultz and O. G. Lopez, Colorado State University, Fort Collins, December 1974.

30. *Determination of Peak Discharge from Rainfall Data for Urbanized Basins, Wichita, Kansas*, C. O. Peek and P. R. Jordon, U.S. Geologic

Survey, Open File Report No. 78-974, Washington, D.C., October
1978.

31. *MITCAT Catchment Simulation Model, Description and Users Manual,
Version 6*, Resource Analysis, Inc., Cambridge, Mass., May 1975.

32. *A Modular Distributed Model of Catchment Dynamics*, Ralph M. Parsons,
Laboratory for Water Resources and Hydrodynamics, Department of
Civil Engineering, Massachusetts Institute of Technology, Cambridge,
December 1970.

33. "Theoretical Frequency Curves," H. A. Foster, *Transactions of the
American Society of Civil Engineers*, Vol. 87, New York, 1924, pp. 142-
203.

34. *A Uniform Technique for Determining Flood Flow Frequencies*, Bulletin
No. 15, Water Resources Council, Washington, D.C., December 1967.

7

Floodplain Delineation

7.1 INTRODUCTION

The purpose of this chapter is to present methods for the determination of
water surface profiles which provide the basis for delineating floodplain areas
for given discharges. These methods allow the engineer to take into consideration channel geometry, channel roughness, obstructions to flow, converg-
ing or diverging flow patterns, and other factors in the delineation of a
floodplain.

7.2 THEORY

For any channel cross section the discharge Q is expressed by

$$Q = VA \tag{7.1}$$

where V is the mean velocity in the channel and A is the cross-sectional area
of the flow normal to the direction of flow. The flow in an open channel is
steady if the depth of flow does not change with time. The flow is *unsteady*
if this depth changes with time.

If in steady flow the discharge is constant throughout the reach of the
channel under consideration, the flow is *continuous*. *Discontinuous* flow is
where water runs into or out of the channel.

It is known in elementary hydraulics that the total energy in any channel
section may be expressed as the total head of water. The total head is equal
to the sum of the elevation above a given datum, the pressure head, and the
velocity head. The total head H at a section is

$$H = Z + d + \frac{V^2}{2g} \tag{7.2}$$

where Z is the elevation of the streambed, d is the depth of flow, and $V^2/2g$
is the velocity head.

Bernoulli Energy Equation

The depths and velocities of flow in a channel, whether an open channel or a
conduit, conform to a principle of the conservation of energy known as
Bernoulli's theorem [1]. This theorem states that the energy of a flow at any
cross section is equal to the energy at a downstream cross section plus the

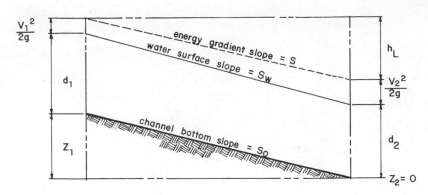

Figure 7.1 Flow in open channels. (Reproduced from Ref. 6.)

intervening losses of energy. As applied to Fig. 7.1, this relationship can be expressed as follows:

$$Z_1 + d_1 + \frac{V_1^2}{2g} = Z_2 + d_2 + \frac{V_2^2}{2g} + h_L \qquad (7.3)$$

where h_L = intervening energy loss

This equation is commonly attributed to the Swiss mathematician Daniel Bernoulli, perhaps to acknowledge his pioneer achievements in hydrodynamics, although it was actually first formulated by Leonhard Euler and later popularized by Julius Weisbach [2].

Losses in Open Channel Flow

The total losses in the Bernoulli equation consist of the friction head h_f, eddy loss, bend loss, bridge loss, and other losses such as at dams and drop structures.

Friction Head
The friction head can be determined by use of Manning's equation:

$$Q = \frac{AR^{2/3}S^{1/2}}{n} \qquad \left(Q = \frac{1.486}{n} AR^{2/3}S^{1/2} \right) \qquad (7.4)$$

where Q = discharge, m^3/sec [ft^3/sec]
  A = cross section of flow area, m^2 [ft^2]
  n = Manning's roughness coefficient (same coefficient for British and metric units)
  $R = \dfrac{\text{area } A}{\text{wetted perimeter } P}$ = hydraulic radius, m [ft]
  S = slope of the energy gradient, m/m [ft/ft]

The value of the roughness coefficient n is discussed later.
  Manning's formula was developed for flow in open channels. It is a special form of Chezy's formula, the derivation of which is contained in most textbooks on elementary fluid mechanics. A nomograph for its solution is presented in Fig. 7.2. The formula can be rewritten as follows:

Table 7.1   Eddy Loss Coefficient K

Flow pattern	Eddy loss coefficient K
Gradually converging flow	0.0-0.1
Gradually diverging flow	0.2
Abruptly converging or diverging flow	0.5

Source: Adapted from *Open Channel Hydraulics* by Ven Te Chow. Copyright 1959 by McGraw-Hill Book Company. Used with the permission of McGraw-Hill Book Company.

$$S = \left(\frac{Qn}{AR^{2/3}}\right)^2 \quad \left[S = \left(\frac{Qn}{1.486AR^{2/3}}\right)^2\right] \tag{7.5}$$

Expressing S as h_f/L, the total friction head for a reach of distance L between two sections can be resolved by the following equation:

$$h_f = L\left(\frac{S_1 + S_2}{2}\right) \tag{7.6}$$

Eddy Loss
Eddy losses originate from the energy losses incidental to velocity head change. At the present time, there are no rational methods available for quantitatively determining these losses. However, it is known that they manifest themselves usually in the form of turbulence. In channel flow where there is an abrupt change in velocity, this energy loss manifests itself in the form of boils and vortexes. This is evident when there is an abrupt decrease in velocity resulting in the conversion of kinetic to static head.

The eddy loss can be expressed as a portion of the change in velocity head, or $K(\Delta V^2/2g)$, where K is a coefficient and $\Delta V^2/2g$ is the change in velocity head. Table 7.1 relates K to various types of flow patterns.

Other Losses
There have been no criteria set for practical evaluation of bend losses. It is recommended that these losses be evaluated by modifying Manning's coefficient as shown in the following material.

The hydraulic conditions at bridge crossings must be carefully analyzed in order to determine bridge losses. It is recommended that the procedures in *Hydraulics of Bridge Waterways* [3] be used to determine the bridge losses.

The evaluation of water surface profiles and energy losses at hydraulic control structures can be extensive. The reader is referred to the many texts available [1,12,13].

7.3 CHANNEL CHARACTERISTICS

The accuracy of any floodplain delineation is dependent on the gathering of data representative of the prevailing hydraulic conditions and the analysis of them through reasonable assumptions and interpretations.

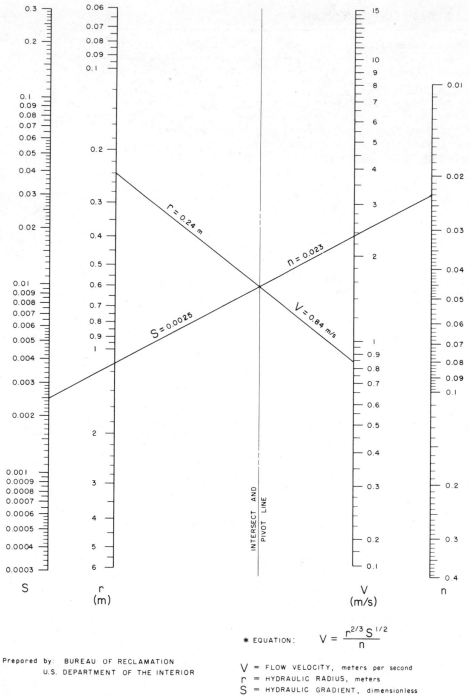

Figure 7.2 Manning formula nomograph. (Prepared by Bureau of Reclamation, U.S. Department of the Interior.)

Channel Cross-Sectional Input

In all cases, cross sections should be taken in a direction perpendicular to the flow lines in the channel. If the cross section is taken from topographic mapping, one must realize that the accuracy of the topographic mapping is probably plus or minus one-half the contour interval. Thus, if the contour interval is 1 m, the actual elevation of any point on the cross section could vary 0.5 m from that shown on the mapping.

If field-surveyed cross sections are used, elevations to the nearest 30 mm (0.1 ft) are usually adequate for cross-sectional data. Horizontal distances to the nearest 300 mm (1 ft) are usually sufficient.

Cross sections should be taken at locations that are representative of the reach to be depicted by that section including all control sections. Cross sections should be spaced closer in areas where the channel cross section varies. Normally, if cross sections are taken at intervals of 150 m (500 ft) in areas of fairly uniform channel cross sections, the resulting floodplain is accurate enough for most planning purposes. The reach length L between sections is usually shown on a plan map, which locates the cross sections.

Floodplain delineation methods that require the separate input of a wetted perimeter should have it inputted at the required accuracy. Cross sections taken from topographic mapping that can be approximated by rectangles, trapezoids, or triangles should have the wetted perimeter calculated by the appropriate geometric formulas. When the wetted perimeter is measured directly from the cross section, the horizontal and vertical scales must be the same.

Channel Roughness

All hydraulic computations of flow in open channels require an evaluation of the roughness characteristics of the channel. Definition of this roughness is subjective and must be developed through experience. It is recommended that the engineer examine the appearance of some typical channels for which the roughness characteristics are known. Color photographs and descriptive data are presented in *Roughness Characteristics of Natural Channels* [4].

Use of Manning's formula requires determination of the roughness coefficient n. The value of n is highly variable and depends on a number of factors. Some of those which exert the greatest influence on the selection of n are listed below:

Surface roughness
Vegetation
Channel irregularity
Channel alignment
Silting and scouring
Obstructions
Size and slope of channel
Stage and discharge
Seasonal changes in vegetation
Suspended material and bed load

Table 7.2 lists average coefficients of roughness for various types of channels. This table is useful for guidance when evaluating selected n values.

Several primary factors affecting the roughness coefficients of channels were considered by Cowan [5], who developed a procedure for estimating the

Table 7.2 Coefficient of Roughness, Average Channels—U.S. Army Corps of Engineers

Value of n	Channel condition
0.025	Small earth channels in good condition, or large earth channels with some growth on banks or scattered cobbles in bed
0.030	Earth channels with considerable growth; natural streams with good alignment, fairly constant section; large floodway channels, well maintained
0.035	Earth channels considerably covered with small growth; cleared but not continuously maintained floodways
0.040-0.050	Mountain streams with clean loose cobbles; rivers with variable section and some vegetation growing in banks; earth channels with thick aquatic growths
0.060-0.075	Rivers with fairly straight alignment and cross section, badly obstructed by small trees, very little underbrush or aquatic growth
0.100	Rivers with irregular alignment and cross section, moderately obstructed by small trees and under-brush; rivers with fairly regular alignment and cross section, heavily obstructed by small trees and underbrush
0.125	Rivers with irregular alignment and cross section, covered with growth of virgin timber and occasional dense patches of bushes and small trees, some logs, and dead fallen trees
0.150-0.200	Rivers with very irregular alignment and cross section, many roots, trees, bushes, large logs, and other drift on bottom, trees continually falling into channel due to bank caving

Source: Reproduced from Ref. 7.

value of n. This procedure is used by the U.S. Soil Conservation Service and others to assist in defining roughness. The method is presented in Table 7.3.

Field Determination
The roughness coefficients can be determined by measuring the water surface profile and rate of discharge in the field for a range of flows. Manning's formula is thus solved for n. This approach to determining the roughness is applicable for most streams. It would result in actual n values for various stages or depths of flow over a period of time as the data were collected. The data would be useful in estimating roughness values on other streams by comparing their physical characteristics with the measured streams.

Sedimentation and Erosion
During a flood a natural channel is subjected to both erosion and deposition. In those instances where the sedimentation process is especially significant,

Table 7.3 A Method of Computing Mean n Value for a Channel

Steps
1. Assume basic n.
2. Select modifying n for roughness or degree of irregularity.
3. Select modifying n for variation in size and shape of cross section.
4. Select modifying n for obstructions such as debris deposits, stumps, exposed roots, and fallen logs.
5. Select modifying n for vegetation.
6. Select modifying n for meandering.
7. Add Items 1 through 6.

Aid for selecting various n values
1. Recommended basic n values

Channels in earth	0.020	Channels in fine gravel	0.024
Channels in rock	0.025	Channels in coarse gravel	0.028

2. Recommended modifying n value for degree of irregularity

Smooth	0.000	Moderate	0.010
Minor	0.005	Severe	0.020

3. Recommended modifying n value for change in size and shape of cross section

Gradual	0.000	Frequent	0.010-0.015
Occasional	0.005		

4. Recommended modifying n value for obstructions such as debris, roots, etc.

Negligible effect	0.000	Appreciable effect	0.020-0.030
Minor effect	0.010-0.015	Severe effect	0.040-0.060

5. Recommended modifying n values for vegetation

Low effect	0.005 0.010	High effect	0.025-0.050
Medium effect	0.010-0.025	Very high effect	0.050-0.100

6. Recommended modifying n value for channel meander

    L_s = straight length of reach L_m = meander length of reach

L_m/L_s	n
1.0-1.2	0.000
1.2-1.5	$0.15 \times n_s$
>1.5	$0.30 \times n_s$

    where n_s = items 1 + 2 + 3 + 4 + 5

Source: Adapted from Ref. 5.

the engineer must consider this process in the hydraulic computations. Erosion during the flood would increase the wetted perimeter, hydraulic radius, and cross-sectional area while deposition would have the opposite effect.

7.4 BASIC COMPUTATIONAL METHODS

Water surface profiles are normally computed by backwater methods. These methods should take into consideration bridge openings, channel drop structures, channel obstructions, and all intervening energy losses. Computations begin at a known point and extend in an upstream direction for subcritical flow. Supercritical flow calculations start upstream and extend in a downstream direction. The plotting of the energy gradient in all floodplain delineations will help to insure against errors.

In the following explanation of various floodplain delineation methods, it is assumed that the flood discharge is already known for the stream segment. Refer to Chaps. 5 and 6 on runoff for an explanation of methods to determine the flood discharge. Only a few floodplain delineation methods are presented here. These are representative of typical water surface profile calculation methods.

Direct Application of Manning's Formula

As discussed, Manning's formula was developed for flow in open channels. This equation determines the normal depth of flow for a continuous flow in a uniform channel. Thus it can be used where these conditions are met. It involves no backwater expressions and should only be used in the preliminary analysis of flows in uniform channels in areas where culverts or other obstructions are absent. It is best used for the determination of the water surface elevation in man-made trapezoidal or rectangular channels. If used on natural channels, it would give only the approximate water surface elevation since the equation does not take into account what is happening to the channel in the adjacent cross sections.

Computational Methods

Manual computational methods are presented below which are suitable for determining the flood profiles.

Direct Step Method
The Direct Step Method is a simple step method applicable to artificial prismatic channels. The channel is divided into short reaches so that the computations can be carried step by step from one end of the reach to the other. There is a great variety of step methods, with some methods being superior to others in certain respects. Reference 1 should be consulted for an explanation of the Direct Step Method.

Modified Standard Step Method
The method described here is the U.S. Bureau of Reclamation "Method B" [3]. Method B is best suited to conditions where flow lengths between sections are different and where the hydraulic elements of n, A, and R vary with the cross section. It is assumed with this method that the elevation-discharge relation is known for the most downstream section.

Figure 7.1 represents the energy balance that is basic to this method. Given the elevation at cross section 1, the water surface elevation at cross section 2 is assumed and the energy equation is checked. If the energy equation does not balance, another water surface elevation is assumed and the iterative process continues until the water surface elevation assumed at cross section 2 checks with the calculated water surface elevation. The calculation of the friction head h_f is of primary importance in this method. This is accomplished by Manning's formula in the form:

$$Q = K_d S_f^{1/2} = \frac{K_d}{L^{1/2}} h_f^{1/2} \tag{7.7}$$

where $S_f = h_f/L$

$$K_d = \frac{AR^{2/3}}{n} \quad \left[K_d = \frac{1.486}{n} AR^{2/3} \right]$$

Solving for h_f,

$$h_f = \left(\frac{Q}{K_d/L^{1/2}} \right)^2$$

This method determines the friction head from the equation

$$h_f = \frac{h_{f_1} + h_{f_2}}{2} \tag{7.8}$$

where h_{f_1} and h_{f_2} are the friction heads computed for the cross sections 1 and 2, respectively.

The total flow at a given section is assumed to be distributed among representative sectional subdivisions of homogeneous character.

The discharge occurring in each sectional subdivision is computed from the formula listed below:

$$Q_s = \frac{K_{d_s}}{L^{1/2}} h_f^{1/2} \tag{7.9}$$

The velocity head h_v is computed by a weighting process where the partial discharges occurring in each subdivision are used. The subdivisional velocity is computed by the equation

$$V_s = \frac{Q_s}{A_s} \tag{7.10}$$

where A_s is the area of the segment. The velocity head is then calculated from the equation

$$h_v = \frac{\Sigma \left(V_s^2 Q_s \right)}{2gQ} \tag{7.11}$$

Eddy losses are calculated from the equation

$$\text{Eddy loss} = M \left(h_{v_1} - h_{v_2} \right) \tag{7.12}$$

where M is taken from Table 7.4.

Table 7.4 Eddy Loss Coefficient M—USBR Method B

Value of $h_{v_1} - h_{v_2}$	M
Negative	-0.5
Positive	0.1

Source: Reproduced from Ref. 3

The use of Method B requires the preparation of various graphs and tables from the collected field data. An example of the step-by-step procedure for using this method follows [6]:

Step 1. Refer to Table 7.5, which is the computational form to be used, and enter the pertinent data in the heading. In this example, the water surface profile is computed for a discharge of 1420 m3/sec

Step 2. Enter the section number of the first two cross sections in column 1. Be sure to skip one or more lines between cross sections and leave room for one line per subsection per cross section. The example shows these entries as 0 and 689.

Step 3. Enter the main channel and overbank reach lengths between sections in column 2 for each subsection. These are noted as 689 and 596 m between sections 0 and 689 for the main channel and overbank reaches as taken from Fig. 7.3, which is the plan map.

Step 4. Enter the known water surface elevation for the downstream cross section in column 3 and the areas of each subsection in column 4. The known water surface elevation for the example is 28.86. The areas of the main channel and overbank flows are taken from a plotted cross section.

Step 5. Calculate K_d for each subsection and enter in column 5. These are equal to 27,533 and 17,118 in the example.

Step 6. Calculate column 6 for each subsection by dividing the individual values of column 5 by the square root of column 2. For the example, the main channel value is computed as follows:

$$\frac{K_d}{L^{1/2}} = \frac{27,533}{(689)^{1/2}} = 1049$$

Step 7. The friction head h_f in column 7 is found by squaring the quantity of the total discharge divided by the summation of the values of $K_d/L^{1/2}$ for all subsections determined in column 6, as follows:

$$h_f = \left(\frac{Q}{\Sigma(K_d/L^{1/2})}\right)^2 = \left(\frac{1420}{1750}\right)^2 = 0.66 \text{ m}$$

Table 7.5 Example Floodplain Delineation

CLIENT Melbourne and Metropolitan Board of Works
JOB TITLE Example Floodplain Delineation
JOB NUMBER 762-29M
COMPUTATIONS BY D.J.L.
CHECKED BY M.A.B.
PHASE 1
DATE 10-11-7_
DATE 12-1-76

WATER SURFACE PROFILE COMPUTATIONS
Example for Q = 1,420 cumecs
STREAM _____
LOCATION _____
REACH _____
SHEET _____ OF _____
PAGE _____

1 SECTION	2 REACH LENGTH L	3 ASSUMED W.S. ELEVATION	4 AREA A	5 $K=\dfrac{AR^{2/3}}{n}$	6 $K_d/L^{1/2}$	7 $\left(\dfrac{Q}{\Sigma\frac{K_d}{L^{1/2}}}\right)^2=h_f$	8 $(K_d/L^{1/2})(h_f)^{1/2}=Q$	9 V	10 V^2Q	11 $h_v=\dfrac{\Sigma V^2 Q}{Q2_g}$	12 $h_{v_1}-h_{v_2}$	13 OTHER LOSSES	14 MEAN h_f	15 TOTAL LOSS	16 ΔH	17 WATER SURFACE ELEVATION
0	689	28.86	315	27,533	1,049		851	2.70	6,204							28.86
	596		1,078	17,118	701	0.66	559	0.53	160	0.23						
					1,750		1,420		6,364							
689	689	29.54	385	30,463	1,161		817	2.12	3,672							
	596		1,268	20,915	857	0.495	503	0.48	139	0.14						
					2,018		1,420		3,811		+0.09	0.01	0.58	0.59	0.68	29.54
689	1,180	29.54	385	30,463	887		813	2.11	3,620							
	995		1,268	20,915	663	0.84	607	0.48	140	0.14						
					1,550		1,420		3,760							
1869	1,180	30.59	433	35,583	1,036		1,296	2.99	11,586							
	995		327	3,119	99	1.565	124	0.38	18	0.42						
					1,135		1,420		11,604		-0.28	0.14	1.20	1.34	1.06	30.60

1/7/69 R.A.F.

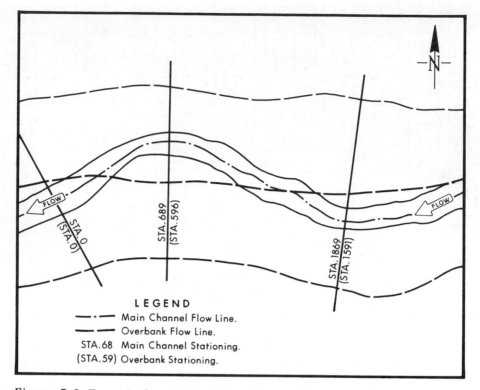

Figure 7.3 Example floodplain delineation plan map. (Reproduced from Ref. 6.)

Step 8. The flow in each subsection is computed by multiplying the value of each individual subsection in column 6 by the square root of column 7. For the example, the overbank area discharge is computed as follows:

$$701 \times (0.66)^{1/2} = 569 \text{ m}^3/\text{sec}$$

Step 9. Calculate the subdivisional velocities by use of the equation $V_s = Q_s/A_s$. Thus, for the main channel,

$$V = \frac{851}{315} = 2.70 \text{ m/sec}$$

Step 10. Square the individual velocities of column 9 and multiply the results by their respective discharges in column 8. For the overbank area,

$$V^2 Q = (0.53)^2 \times 569 = 160$$

Step 11. Calculate the weighted velocity head by dividing column 10 by the product of the total discharge multiplied times $2g(2 \times 9.81)$ m/sec. Thus, in the example

$$h_v = \frac{\Sigma(V^2 Q)}{Q\,2g} = \frac{6364}{1420 \times 19.6} = 0.23 \text{ m}$$

Step 12. Return to step 1, if this is the first section of an iteration. Assume a water surface elevation for the next upstream section. Perform similar calculations through column 11 as described above. If this is not the first section of an iteration, calculate the algebraic difference in velocity heads and enter in column 12; thus, in the example:

$$h_{v_1} - h_{v_2} = 0.23 - 0.14 = +0.09 \text{ m}$$

Step 13. Enter the eddy loss in column 13. The loss can be determined by using Table 7.4 and the equation

$$\text{Eddy loss} = M \left(h_{v_1} - h_{v_2} \right)$$

Record the result in column 13. In the example, this amounts to

$$0.1 \times 0.09 = 0.01 \text{ m}$$

Step 14. Column 14 is the mean friction head loss h_f between sections. Calculate this value by averaging the values in column 7. Thus, in the example,

$$h_f = \frac{0.66 + 0.495}{2} = 0.58 \text{ m}$$

Step 15. The total loss in column 15 is calculated by adding columns 13 and 14.

Step 16. Column 16 is the algebraic sum of columns 12 and 15.

Step 17. The water surface elevation in column 17 is determined by adding column 16 to the water surface elevation of the downstream section. In this example, it is computed as follows: $28.86 + 0.68 = 29.54$. If the water surface elevation in column 17 differs from the assumed water surface elevation in column 3 by more than 30 mm (0.1 ft), another water surface elevation should be assumed and the calculation repeated. If this difference is less than 30 mm, the cycle is complete and a water surface at the next upstream section is assumed and the step procedure is repeated [3].

Computer Programs

There are numerous computer backwater programs in existence. The procedure described here is the HEC-2 backwater program developed by the U.S. Army Corps of Engineers. A detailed description of the program is summarized in the *HEC-2 Water Surface Profiles Users Manual* [7]. The data base required for this program consists of cross-sectional channel shapes, roughness, and details on culverts and bridges.

This program is a modification of program 723-G2-L214A, developed by the U.S. Army Corps of Engineers [7]. The input requirements have been modified to allow the use of additional options, to provide for future expansion of the program, and to simply input preparation.

The program computes and plots (by printer) the water surface profile for river channels of any cross section for either subcritical or supercritical flow conditions. The effects of various hydraulic structures such as bridges, culverts, weirs, embankments, and dams may be considered in the computation (the user should be very careful with application of bridge routines and verify results.) The principal use of the program is for determining profiles for various frequency floods for both natural and modified conditions. The latter may include channel improvements, levees, and floodways. Input may be in either British or metric units.

The computational procedure is similar to Method 1, U.S. Army Corps of Engineers [3]. This method applies Bernoulli's theorem for the total energy at each cross section and Manning's formula for the friction head loss between cross sections. In the program, average friction slope for a reach between two cross sections is determined in terms of the average of the conveyances at the two ends of the reach. Other losses are computed using one of several methods. The critical water surface elevation corresponding to the minimum specific energy is computed using an iterative process.

The computation begins at a control section (location of known water surface elevation) in the river channel and proceeds upstream for subcritical flow or downstream for supercritical flow. In cases where flow passes from subcritical to supercritical, or vice versa, during computations, it is necessary to compute the entire profile twice assuming alternately subcritical and supercritical flow. From the above results the most likely water surface profile can be determined.

The water surface elevation for the beginning cross section may be specified as critical depth, as a known elevation, or by the slope area method.

7.5 SPECIAL PROBLEMS

During the process of delineating a floodplain, special problems often arise. A discussion of those items which most frequently arise follows.

Stage-Discharge Relationship for Initial Section

Most of the discussions so far have assumed that the elevation of the down-stream water surface for a given discharge is known. This is not always true.

For reservoir backwater computations, the reservoir water surface elevation is the starting elevation. Whether it be the top of the conservation pool, top of the flood control pool, the maximum water surface, etc. depends upon the purpose of the backwater profiles. The starting point of the back-water curve within the reservoir for any given pool elevation may be selected by applying the criterion that the velocity head h_v is less than 15 mm (0.05 ft) through the cross-sectional area. The average velocity then should be 0.55 m/sec (1.80 ft/sec) or less [3].

The starting elevation for computing the profiles of tailwater elevations below dams can sometimes be found by using the stage-discharge relationship for a stream-gaging station if it is located a short distance below the dam.

Where there is no stage-discharge relationship available, such a curve must be developed. Extreme accuracy of this relationship is not important if several section computations are made before progressing into the area where greater accuracy of the profiles is required. The computed profiles for a given discharge will usually converge after three or four sections even though the starting elevations vary by as much as 0.9 m (3 ft).

A method in the development of a stage-discharge relationship is the application of a constant slope to the conveyance of the section. This slope may be that of an observed high- or low-water profile surveyed from high-water marks. Sometimes the slope of the streambed profile is used for a reach through the section. The section conveyance is calculated by the equation

$$K_d = \frac{AR^{2/3}}{n} \qquad \left[K_d = \frac{1.486}{n} AR^{2/3} \right] \qquad (7.13)$$

where $Q = K_d S^{1/2}$

Control sections with flow at critical depths make excellent starting sections, since the stage-discharge relationship is well established by the critical depth determination.

Critical Flow Conditions

Critical flow conditions occur where the channel slope steepens abruptly, at a natural construction in the channel, and at locations where the placement of structures raises the channel bottom or constricts the channel width. Critical conditions at a section are reflected in the computations when a balance cannot be reached in Bernoulli's equation. Before proceeding with the computations, it is necessary to determine the critical depth or elevation for the section.

There are a number of ways for checking the existence of critical or subcritical depth. The basic equation for flow at critical depth is:

$$\frac{Q^2}{g} = \frac{A^3}{T} \qquad (7.14)$$

where Q = discharge
       g = acceleration of gravity
      A = area
      T = water surface width

This equation must be satisfied when flow is at critical depth.

A preliminary check can be made of critical conditions by determining the critical velocity from the following equation, which results from an algebraic transformation of Eq. (7.14), letting Q = VA:

$$V_{cr} = 3.13 \left(\frac{A}{T} \right)^{1/2} \qquad \left[V_{cr} = 5.67 \left(\frac{A}{T} \right)^{1/2} \right] \qquad (7.15)$$

Thus, when the computed velocity exceeds critical velocity V_{cr} it can be assumed that critical flow exists.

Other means can also be used in determining critical flow; however, they will not be discussed since all are based on the various ways of manipulating the basic equation with the data at hand. The important point of the analysis is that the critical depth elevation is used to proceed with computations when the critical discharge is less than the computed discharge. With natural stream conditions supercritical flow will not usually occur for any appreciable distance.

Bridge Openings

Bridge crossings are often crucial factors in the computations of water surface profiles. Establishing cross sections at the bridge and at 15 and 30 m upstream and downstream from the bridge greatly facilitates the backwater computations in order to obtain realistic results. If cross sections have not been established, however, the hydraulic properties of the adjacent upstream and downstream sections are assumed to apply.

If the bridge becomes a submerged obstruction at higher flood flows, backwater elevations can be determined by computing flows by the orifice formula $Q = CA\sqrt{2gH}$ and the broad-crested weir equation $Q = CLH^{3/2}$, through the submerged bridge opening and over the roadway and bridge, respectively. The *Hydraulics of Bridge Waterways* manual [3] contains methods for determining backwater curves through bridge sections. The engineer must make a thorough investigation of the hydraulics involved at the bridge crossing and then based on judgment select the method which will give the best practical solution to the problem.

Comparison with Recorded High-Water Marks

When an inconsistency between a calculated floodplain and historic high-water marks is found, the calculations should be reviewed in detail to determine the reason for the difference. If the computed floodplain is correct, then the analysis of the difference should be explained in the report.

Overland Flow

Quite often in flood studies, water flows at a shallow depth in overland flow or it overtops a ridge, road embankment, or berm and is trapped for a distance or diverted before it can return to the main channel. In cases like this, the trapped water may not be hydraulically connected to the main channel flows.

A careful analysis must be performed to determine how much water is flowing overland or is trapped out of the main channel and a separate backwater analysis of this water must be performed.

7.6 PROCEDURE FOR DELINEATING FLOODPLAINS

The process of delineating a floodplain may be divided into three main categories: obtaining and processing field data, computation of the water surface profile, and compilation of a written report substantiating the resultant floodplain. Care must be taken in each of these categories to insure that all work pertaining to these items is carried out in a professional manner to insure the accuracy of the delineated floodplain.

Obtaining Field Data

The methods discussed in this book require determination of the hydraulic parameters of n, A, R, and L, as well as a general analysis of the situation for the most practical approach. Generating the field information can be very costly. It is recommended that an early reconnaissance be made of the area by a hydraulic engineer to determine the extent of the necessary data. When the need for data has been established, the cost of the field surveys can often be reduced by integrating the surveys with available drawings and site topography.

A critical step involves locating the sections for survey. Aerial photographs and topographic maps will aid in the location of sections to be surveyed. Sections should be located so the average area, hydraulic radius, and n values of the section will be representative of the reach length. Channels that are alternately wide and narrow should have sections located in the narrow and wide reaches so the average condition can be defined. Sections should also be located to reflect the influence of grade changes and section controls such as bridge crossings.

In general, the distance between cross sections can be longer for channels with flat slopes. Normally, if cross sections are taken at intervals of 150 m (500 ft) in areas with control sections and at fairly uniform channel cross sections, the resulting floodplain is accurate enough for most planning purposes. Other considerations include the size of the channel and the discharges being investigated. Distances can be greater than those stated above for large channels and higher discharges.

When the criteria for the section survey are set, the hydraulic engineer should first determine the maximum discharge for which the profiles will have to be computed. The required cross-sectional area can be approximated by using a rule of thumb that typical flood velocities in nature are up to 2.4 m/sec (8 ft/sec). This will help in establishing the detail of the survey and give some idea of the elevation to which the section surveys should be carried. Also, as a general rule, the main channel has the largest conveyance within a section; therefore, definition of this segment should be good. It is often necessary to take soundings in the deeper-flowing portions of the channel. A topographic map may not define a section adequately for realistic profile computations, particularly in the main channel portion; hence, it is sometimes recommended that field surveys be made. It is good practice to record in field survey notes the section stationing at the points of changing n values. If the hydraulic engineer will go over the sections in the field with the surveyor, the sections can be flagged for changing n values. All sections should be located transversely to the bulk of the flow at the section. When the sections are surveyed, maximum high-water marks should be recorded wherever possible as well as the time and elevation of main channel flow. The maximum high-water marks can aid the engineer in the determination of Manning's n if the peak discharge of the channel is known for the event that produced the high-water mark. The discharge can be secured from the records of nearby gaging stations, if available. The recorded high-water mark will help relate the delineated floodplain to a recorded historic flood.

The n value should receive careful consideration since it influences the computations by the first power as contrasted with the two-thirds power of the hydraulic radius and the one-half power of the slope in Manning's formula. It is possible to assign an average n value for a section, but this is extremely difficult for all stages considering the changing relationship of discharge between the various segments of the section. It is easier to assign the n values to segments of the total section. Main channel sections may have an increasing n value for higher stages due to overhanging vegetation.

The cross-section surveys and assignments of n values define the parameters of n, A, and R, leaving the length parameter L to be determined. The sections normally are tied into a traverse so a plan map of the section layout can be developed. The main channel and overbank areas can be developed at each section, but this does not necessarily determine the meander length of the channel nor the path of the overbank flow.

There are three methods for obtaining this data. If topographic maps or even reliable plan maps are available, the section plan may be plotted on these maps and channel lengths measured. A channel plan map may be developed

from aerial photographs upon which the cross section plan may be plotted
and thus the flow lengths determined [11]. If aerial photographs and reliable
maps are not available, then measurements must be taken in the course of the
traverse survey to produce the extent of the overbank areas and the main
channel location from which a plan map may be developed. Unless the traverse completely represents the channel alignment, this length should not be
used in the determination of L.

Photographs representative of the typical section conditions provide
valuable information to the hydraulic engineer, especially if n value adjustments are necessary. One or two water surface profiles of high-water marks
through the reach are valuable aids in determination of n values and will add
to the accuracy of the profiles.

Processing Field Data

If detailed topographic mapping is not available, the survey notes, after the
routine reduction, will be the basis of developing a plan map, cross-section
plots, streambed profile, and table of reach lengths.

The plan map, as discussed in the preceding section, can be developed
in one of the three following ways: (1) overlaying the traverse and section
plot on a plan map or topographic map; (2) overlaying the traverse and section plot on a plan map developed from aerial photographs; or (3) development
of the plan completely from the survey notes. The first two methods are ways
of reducing the cost of the fieldwork.

For manual computational methods cross sections should be plotted next.
The scales chosen should be such that a realistic concept of the section can be
obtained. For computer methods, the field data may be converted directly
into computer input. Information from the field notes regarding n values,
low-water elevations, high-water marks, and descriptions should be included
in the data input.

In manual computational methods the area and hydraulic radius computations will be made from the cross-sectional plots; hence, the scale of the plots
should be chosen relative to the methods used for computation. A and R
computations are commonly made by planimetering the area and measuring the
wetted perimeter or by computing both parameters. If the wetted perimeter
is determined by measurement, the same scale must be used for both the
abscissa and ordinate.

The sectional parameter of area and hydraulic radius for use in manual
computations are best handled by drawing curves of the area and hydraulic
radius vs. elevation. A few determinations of the A and R for different
elevations will make it possible to draw a reasonably defined curve.

From the plan map, the distance L between sections can be determined.
This distance for a large channel is merely the channel length between sections. If there are large overbank flows with significant differences in the
flow lengths between the overbank and main channel, then all these lengths
should be determined.

For both manual computational and computer methods, a profile should be
plotted. The horizontal stationing is determined from the reach length between sections and the streambed elevation taken at the lowest point in the
cross section from the cross-sectioned information. All observed water surface and high-water mark elevations should be included on this plot. The
scales for this plot should be extended to enable a later plot of the computed
profiles. This profile should be considered a working draft; hence, it is

recommended that the scales be large enough to obtain a reasonable concept of the profile characteristics. A final reduced profile for a finished report can be made later from this working draft.

Computation of Water Surface Profile

Water surface profiles should be determined by one of the methods described or by another generally accepted procedure. The chosen procedure should take into consideration bridge openings, channel drop structures, channel obstructions, and all intervening losses in the computation of the water surface profile.

After the water surface profile is determined, the resulting floodplain limits should be plotted on topographic mapping. This is accomplished by marking the floodplain top width at each cross section on the topographic mapping. The floodplain is then drawn by connecting top width marks on the cross sections. Care must be taken to insure that all elevations of the water surface profile correspond to the elevation of the floodplain between cross sections. The drawn limits of the floodplain should cross contours of the same elevation only once; locations where contours of the same elevation are crossed should normally be directed opposite each other across the channel.

Compilation of Data

A report should be prepared for each floodplain delineation study. The method and computational procedure used in the delineation of the floodplain should be described.

The following items are recommended for inclusion in each study:

A topographic map (if available) showing traverse and layout of the cross sections should be included. Areal distribution should be shown of vegetation, such as brush, trees, and grass, and type of land use—cultivated, grazing, or barren. For backwater studies, this map will be very helpful in showing the extent of inundated areas for the specific discharge computed.

Land and (if available) aerial photographs should be included in the report. Notes are necessary for describing unusual hydraulic conditions of the reach.

High-water marks for any observed discharge should be plotted.

When a gaging station is not located within or near the reach under study, a stage-discharge rating curve for the lowermost section will need to be established.

A plotting of the cross sections should be furnished along with a tabulation of the pertinent hydraulic properties.

Distances between sections should be tabulated.

A graph is needed showing the water surface profile and energy profile for each computed discharge. The bed profile should also be included.

Care must be taken in the preparation of the data to insure that all assumptions used in the delineation of the floodplain are clearly stated. These assumptions could include, but are not limited to:

Debris blockage of bridge openings and channel sections
Sedimentation relation to channel geometry
Washout of bridges, berms, dikes, roads, and related facilities

In all cases, the floodplain delineation data should be accompanied by a supplemental hydrological study that shows how the flood frequency vs. discharge relationship was defined.

7.7 CONTINUED DEVELOPMENT IN FLOODPLAIN MAPPING

Flood Insurance Rate Mapping System

Surveying the basic river-floodplain cross sections and drafting the flood profile maps represent the larger component of cost in performing a flood study, while analysis of water-discharge computations, computer runs, profile analysis, and floodway determinations represent the smaller component of the total cost of the project.

The Flood Insurance Rate Mapping System (FIRMS) concept was designed jointly by the Jet Propulsion Laboratory and Xerox Electro Optical Systems for high-speed surveying of floodplain elevations [8]. Data from an airborne crosspath scanning laser rangefinder are reduced using image-processing techniques to produce elevation maps. It is anticipated that aircraft operating at 1000 to 2000 ft above ground level could survey 200 mi^2/day. The elevation maps can be overlaid on existing maps.

FIRMS presents an integrated approach to the surveying, modeling, and mapping components for detailed flood studies. The instrumentation and computer software required to produce digital tape data sets of elevation, contour maps, and orthophotoquads (i.e., topographically correct photography) of flood-prone communities are designed to interface directly with existing flood hydrology models. FIRMS will, therefore, reduce costs precisely where they are greatest at present, that is, in the mapping.

Aerial Profiling of Terrain System

Another high-speed method of floodplain profiling, the Airborne Profiling of Terrain System (APTS), has been designed by the Charles Stark Draper Laboratories, Inc. under the sponsorship of the U.S. Geological Survey. This is an airborne instrument package capable of providing stream-valley profile data to current accuracy standards and at the same time accomplishing substantial savings in fieldwork, time, and costs.

The stream-valley profiling system is designed to be carried by a relatively light, twin-engine aircraft. The performance goal of the system is profiling of stream-valley geometry to an accuracy of ±15 cm in the vertical coordinate and ±3 m in the horizontal coordinates. Profiles are required at stream cross sections spaced, say, 1/2 km apart throughout a nominal survey area 3 km wide by 30 km long. The airborne instrument package consists of a laser profiler, a three-gimballed inertial platform, and a two-axis laser tracker. The tracker measures distances and directions to previously surveyed retroreflectors to provide update information. The inertial platform and laser tracker together constitute a high-accuracy, three-coordinate position datum which in effect enables the laser profiler to measure the terrain elevation [10].

Auto Surveyor

A mapping system, called the Auto Surveyor, was produced by Span International, Inc. and perfected by the U.S. Army Corps of Engineers. The Auto Surveyor uses an inertial system to measure accelerations along three axes. A laser geodometer can be coupled to the inertial system to increase the system's capability. The system is moved from point to point in a van or helicopter.

The Auto Surveyor coupled with the laser geodometer is useful to survey a large area quickly, and has been used by the Corps in Portland, Oregon to provide ground control points for aerial photographic mapping.

ERTS-1

The Landsat Satellite Program (Earth Resources Technology Satellite) has been used for identifying floodplain features [9]. Each Landsat experimental satellite is in a circular, near-polar, sun-synchronous orbit so that the same point on the Earth's surface is viewed every 18 days at the same time of day. The orbit is at an altitude of about 920 km (about 570 mi). The satellite can record a 115 by 115 nautical mile area. Flood-prone areas can be mapped through photointerpretation but at scales no larger than 1:250,000 (1 in. = 4 mi). This allows floodplain mapping to be useful for regional planning, but it is not suitable for local applications. This method of floodplain delineation would be useful for determining floodplain areas that should be avoided in the construction of new industry or planned communities. Landsat was used in 1973 to rapidly map the extent of the flooding in the Mississippi River Valley and could be used in future floods to record the maximum extent of flooding even after flood waters have receded. Data dissemination from the Landsat program is managed by the U.S. Geological Survey.

REFERENCES

1. *Open Channel Hydraulics*, Ven Te Chow, McGraw-Hill, New York, 1959.
2. *Engineering Fluid Mechanics*, Charles Jaeger, translated from German by P. O. Wolf, Blackie, London and Glasgow, 1956.
3. *Hydraulics of Bridge Waterways*, U.S. Department of Transportation, Federal Highway Administration, Bureau of Public Roads, Office of Engineering and Operations, Bridge Division, Hydraulic Branch, Washington, D.C., 1970.
4. *Roughness Characteristics of Natural Channels*, U.S. Department of the Interior, Geological Survey Water Supply Paper 1849, Washington, D.C., 1967.
5. "Estimating Hydraulic Roughness Coefficients," Woody L. Cowan, *Agricultural Engineering*, Vol. 37, No. 7, July 1956.
6. *Guide for Computing Water Surface Profiles*, U.S. Department of the Interior, Bureau of Reclamation, Sedimentation Section Report, Washington, D.C., November 1957.
7. *HEC-2 Water Surface Profiles Users Manual*, Hydrologic Engineering Center, U.S. Army Corps of Engineers, 609 Second Street, Davis, Calif., 1973.
8. *FIRMS, Flood Insurance Rate Mapping System*, Vol. 1, Jet Propulsion Laboratory, California Institute of Technology, Pasadena, December 1976.

9.   *Third Earth Resources Technology Satellite Symposium*, Vol. III, National Aeronautics and Space Administration, Washington, D.C., 1974.

10. *Aerial Profiling of Terrain*, J. W. Hursh, G. Mamon, and J. A. Soltz, Charles Stark Draper Laboratory, Cambridge, Mass. 1977.

11. "Satellites Helping Solve Down-to-Earth Civil Engineering Problems," *Civil Engineering*, August 1975; pp. 49-53.

12. *Design of Small Dams*, U.S. Department of the Interior, Bureau of Reclamation, Washington, D.C., 1973.

13. *Engineering Hydraulics*, Hunter Rouse, Wiley, New York, 1950.

8

Nonstructural Planning

8.1 FLOODPLAIN MANAGEMENT

Floodplain management includes all measures for planning and action which are needed to determine, implement, revise, and update comprehensive plans for the wise use of floodplain lands and their related water resources. This includes both preventive and corrective actions, the latter encompassing flood control works such as storage basins and channelization along with flood proofing and land use changes.

Structural measures affect floodwaters while nonstructural measures affect activities in the floodplain. Because of their relatively recent acceptance as a viable alternative, nonstructural measures are discussed separately. This is done to increase awareness of the nonstructural approach.

Land use management is the keystone of nonstructural floodplain management. In its broadest terms it involves both land use and runoff controls. Land use management is used in this chapter to describe policies of land management for prudent and productive use of hazardous areas. It involves a set of actions by all levels of government in cooperation with the private sector which can be relied upon to guide the wise use of public and private floodprone land. These actions include acquisitions, legislative controls, land valuations, investments, and the provision of information.

Early Examples

Nonstructural floodplain management is not new. For instance, Empress Maria-Theresa (1740-1780) of the Austro-Hungarian Empire decreed that a wide floodplain be left undeveloped adjacent to Vienna for the Danube River. Current public works engineers for Vienna are grateful for her foresight. Great floodplains are left undeveloped and undiked for 50 km upstream from Vienna to temporarily store flood peaks, thus further protecting Vienna even today by providing floodplain storage.

Ancient Egyptians managed the Nile floodplain for agricultural enhancement through the entire length of the country. Management included an early warning and flood forecast system upstream of Aswan at the first cataract.

U.S. Development

Modern articulation of floodplain management in the United States dates to 1945 when Dr. Gilbert White prepared his paper entitled *Human Adjustment to*

Floods: A Geographical Approach to the Flood Problem in the United States
[1]. Progress was slow until 1966 when the Task Force on Federal Flood Control Policy, chaired by Dr. White, reported to the President of the United
States in a document entitled *A Unified National Program for Managing Flood
Losses* [2].

U.S. Floodplain management policy has evolved under four presidents.
It is aimed at reducing national flood losses. In 1966 President Johnson
issued Executive Order 11296, which signaled a major change from the previous strategy of structural solutions to flood problems. The Flood Insurance
Act followed, along with the Flood Disaster Protection Act (PL 93-288) under
President Nixon. President Ford was a strong supporter of the Johnson-
Nixon flood efforts and drew up Executive Order 11910. In 1977 President
Carter issued Executive Order 11988 and instructed federal agencies to make
it easier for local governments to acquire floodplain lands in a directive dated
July 12, 1978, which also further stressed the policies of the three former
Presidents. A federal policy to respect and adjust rather than change nature
has been adopted. The federal floodplain management policy is based upon
acknowledging the forces of floodwaters and the prescriptive easement that
floods have in the low-lying natural flood paths.

Implementation of the national policy is achieved via a direct partnership
between the federal government and local governments. Cooperation is
achieved from the local governments using numerous incentives, the primary
one being the National Flood Insurance Program which requires cooperation
for a city or county to be eligible for benefits. Another incentive relates to
distribution of federal grants and assistance following an actual flood disaster.

However, the states are not silent bystanders in implementing the national nonstructural policy. Colorado, for example, has a statewide model floodplain regulation which sets minimum standards for local government. Governor Lamm of Colorado issued a series of Executive orders which established
state policy similar to that at the federal level.

8.2 APPLICABILITY

Nonstructural floodplain management measures can be applied to existing
development, future development, and heavily-encroached-upon floodplains.
Each condition met in the field requires a specially developed combination of
methods. Existing development and heavily encroached floodplains require
corrective actions. When a nonstructural approach is programmed, it may be
supplemented with structural measures to achieve the desired degree of protection. However, the nonstructural measures can move independently of the
supplementary structural measures. Early implementation of the nonstructural measures will help achieve a reduction in flood hazard without waiting for
the balance of the program.

Unfortunately, some engineering firms and public works engineers are
oriented toward the structural approach to flood management. This is understandable because of the orientation of their training toward building and
structures rather than land use planning. It is for this reason that nonstructural floodplain management consideration and its application must be
imposed as governmental policy.

Examples of recalcitrant local governments ignoring the benefits of floodplain management based on nonstructural methods abound. Yet this does not
affect its applicability; it only serves to emphasize the need for understanding
and education along with the need for strong and clear national policy.

Following a devastating flood, such as those in Rapid City, South Dakota in 1972 and the Big Thompson Canyon flood in Colorado in 1976, the opportunities for nonstructural floodplain management are greatest. Yet only state and national intervention through a combination of regulations and financial assistance directed these two recovery efforts toward nonstructural methods rather than merely rebuilding what was there previously.

8.3 PREVENTIVE ACTIONS

The preventive actions available to a community to reduce flood losses include a wide variety of measures. These include:

> Control of floodplain land uses
> > Floodplain regulations
> > Floodplain acquisition
> > Subdivision regulations
> > Control of water and sewer extensions
> Floodplain information and education
> Flood forecasts and emergency measures
> Flood proofing
> Flood insurance

When preventive measures are implemented, a range of benefits can be realized. These include the following:

> Reduced flood damages
> Reduced urban disruption and social chaos
> Reduced street construction costs
> Reduced street and highway maintenance costs
> Water-oriented recreation opportunities
> Multipurpose open space
> Less problems with high water table
> Trails for hiking, biking, and horseback riding
> Preservation of natural floodplain storage
> Reduced sewer pipe infiltration and inflow
> Improved stream water quality
> Better quality of nonpoint discharges of urban runoff

8.4 CORRECTIVE ACTIONS

To solve our inherited flood problems and spare future generations from them, corrective actions must be taken. For nonstructural solutions, these actions fall under the general headings of flood proofing and land use adjustments. In a comprehensive program, preventive actions need to be integrated with corrective actions. This allows a dynamic interaction between elements to achieve the most cost-effective design.

Flood Proofing

Flood proofing is a means of adjusting an existing structure to the flood hazard by making facility modifications to reduce the damage potential during flood periods. Flood proofing is discussed later.

Land Use Adjustments

Corrective measures include land use adjustments such as relocation of structures, programmed removal of noncompatible structures, and purchase of
floodplain properties. Properties purchased may be leased back for temporary use with scheduled razing in the future. Urban renewal projects may be
used to expedite adjustments in land use.

Nonconforming Uses

When land is rezoned, the existing use of the land or of the buildings thereon
may no longer conform to the zoning now in effect. A change in building
regulations has the same effect on existing buildings. Such use or building
is then held to be *nonconforming*. The general policy is that such uses and
buildings are permitted to continue, but that nonconformity should eventually
cease.

8.5 NONSTRUCTURAL PLAN COMPONENTS

A nonstructural floodplain management plan must call upon the use of many
components to be effective and practical. It is more demanding upon the
engineers and planners engaged in flood control and it requires a high degree
of performance and professionalism. On the other hand, many nonstructural
measures can be implemented without heavy capital expenditures.

The components of a nonstructural program fall under the broad categories of land use management, early warning, land runoff control, flood
proofing, insurance, and relief and rehabilitation. The components are described below along with the type of basic data needed and the important
flood-prone area maps.

Control and Floodplain Land Uses

Floodplain land uses may be affected by either floodplain controls or land
acquisition. By controlling the amount and type of economic and social growth
in the floodplain, flood losses are reduced and net benefits from suitable
floodplain use increase.

Although this approach may require long periods to exercise its full
effect, the incentives to communities are strong. They include reduction in
the exposure to risk, reduction in public costs for relief and rehabilitation,
and decreased dependence upon protective works.

Methods used include land use controls, land acquisitions, subdivision
regulations, and control of water and sewer extensions.

Floodplain Regulations
The delineation of a flood-prone area can be undertaken as described previously. The runoff should be based upon future development of the basins in
accordance with the areawide planning proposals for future development, and
with floodplains and watercourses generally in their existing condition.

The floodplain delineation and flood-prone area designation will recognize
nature's prescribed and natural easement along each of the watercourses.
Land uses within the floodplain areas will be controlled in accordance with
relevant provisions of the planning scheme. The following chapter describes
floodplain regulation in more detail.

Floodplain Acquisition
Floodplain land can be publicly acquired by purchase or by the dedication of
such lands by developers as permanent drainage reserves or open space.

Acquisition of floodplain land may be a cost-effective form of floodplain
management. When it is done, other benefits may be realized.

Floodplain storage will be preserved. Open space for selected active
and passive recreation may be provided. Space for car parks may be provided. Trails may be constructed along the floodplains to link major recreational and park areas. Finally, floodplain ecosystems may be preserved.

Subdivision Regulation
The subdivision regulations can be used to prevent the creation of development within floodplains and to regulate and control stormwater runoff. Proper planning will allow the preservation and/or restoration of natural and
beneficial floodplain values.

Control of Water and Sewer Extensions
Many cities are in the advantageous position of controlling not only drainage,
but also water and sewer utilities.

To protect the public health, safety, and welfare, as well as the public
investments in the utility lines, water and sewer services generally should not
be placed in designated floodplains.

As a pollution control step, installation of sewer lines in flood-prone
areas should not, in general, be allowed. This will mitigate the infiltration
and inflow of floodwater into the sewer lines and, in turn, reduce the probability of sewer overflows which have the potential of causing sewers to back
up into homes or discharge raw sewage into watercourses.

Control of water and sewer extensions, coupled with the town planning
functions, allows the city to actively discourage the unwise use of floodplains.

Floodplain Information and Education

As part of a long-term program, the delineation of the flood hazard should be
undertaken. Information on the flood hazard is a vital input to any flood
damage reduction action.

The development of floodplain information should be accompanied with a
public awareness program. Floodplain maps accompanied by information on
measures which a property owner can undertake to mitigate his potential
losses can be provided to public agencies and landowners when available. The
news media can be contacted and encouraged to disseminate information on
floodplains and adjustments. Pamphlets can be published that contain mapping
so that citizens can understand the areal extent of the floodplain and potential
adjustments which can be undertaken.

Significant educational efforts will help the citizens to become aware of
the range of actions which can be taken to mitigate the effects of flooding.
Films, videotapes, and slides can be used to help communicate this information.

Various methods can be used to mark and make known the floodplain in
the field such as showing the 1 percent flood level on public buildings and
at bridges and by using other easily recognized techniques for providing
warning signs.

Public and private financing institutions should be drawn into the floodplain management process. The private institutions should be provided with
informational packets relating to the flood hazard and appropriate adjustments.

Flood Forecasts and Emergency Measures

Prediction of floods with appropriate forecasts or warnings delivered with
credibility can trigger a series of emergency measures which will reduce flood
damages. This is true even where only a short lead time is available. A
simple warning can significantly reduce the tragedy of a flood.

Forecast and warning of a flood can be very effective when sophisticated
communication techniques are used to quickly assess information on upstream
flows and basin rainfall patterns. When such data are available, hydrologists
will be able to predict the level of anticipated flooding at downstream locations
so that warnings could be given with 3 to 18 hours of lead time.

Use of radar scanning for cloud buildup can provide additional warning
time for public works officials and emergency personnel to be on standby.

A flood forecasting system will reduce flood losses as it stimulates
appropriate emergency actions before the floodwaters reach vulnerable areas.
People, equipment, and materials should be moved from floodplain sites to
locations above anticipated flood heights. Equipment that cannot be moved
should be treated to mitigate water damage. Where flood-proofing measures
have been incorporated into structures, they should be activated, e.g., bulkheads secured in place, in anticipation of the flood. Adjustments should be
made to utilities to assure continuation of vital services. Equipment and
personnel should be dispatched to critical channel constrictions, including
bridge openings, and control works at storage basins to remove debris to
assure proper functioning. In essence, a disaster preparedness plan must
be developed by appropriate government agencies to assure that proper
emergency measures are implemented in an effective and timely manner.

Flood Proofing

Flood proofing is an important nonstructural floodplain management tool.
Flood proofing consists of those adjustments to structures and building contents which are designed or adapted primarily to reduce flood damages.
Flood proofing is applicable mainly to substantial structures such as many
which are used for commercial and industrial purposes. This concept allows
private property managers to take actions to reduce their flood risks.

Components

Many adjustments can be made to structures and contents which will mitigate
the effects of flooding [1-6].

Flood proofing should be carried out under the direction of a professional engineer or architect. Figure 8.1 depicts a range of measures for a typical
building.

General measures which may be mutually exclusive are:

Anchorage to resist flotation and lateral movement (inasmuch as floating
    residential homes are a hazard to property and contribute significant-
    ly to the debris problem downstream)
Watertight doors, bulkheads, shutters, and sandbags
Reinforcement of walls to resist collapse from hydraulic pressure
Waterproofing walls to control seepage
Addition of mass to resist flotation
Installation of pumps to control seepage
Check valves on sewerage and stormwater drains
Reduction or management of water table to relieve hydraulic pressures

Explanation:
1. Permanent closure of opening with masonry
2. Thoroseal coating to reduce seepage
3. Valve on sewer line
4. Underpinning
5. Instrument panel raised above expected flood level
6. Machinery protected with polyethylene covering
7. Strips of polyethylene between layers of cartons
8. Underground storage tank properly anchored
9. Cracks sealed with hydraulic cement
10. Rescheduling has emptied the loading dock
11. Steel bulkheads for doorways
12. Sump pump and drain to eject seepage

Figure 8.1 A flood-proofed structure. (*Source*: Ref. 3.)

> Raising of electrical control panels
> Protection of sewer manholes from entrance of floodwaters
> Distribution of contents stored within structures

Building Code
Building codes should be modified to include flood-proofing measures [3] to mitigate existing flood hazards. Technical information regarding flood proofing can be prepared and disseminated through a floodplain information and education program.

Flood Insurance

The U.S. federal government has a workable flood insurance program which balances public subsidies with requirements for appropriate land use controls. Existing flood-prone properties can obtain subsidized insurance. New buildings are subject to actuarial rates. The U.S. program has shown that actuarial rates virtually rule out any further development of floodplains. Experience has shown that the benefits derived from a floodplain site cannot offset the costs associated with the flood risk if the owner must absorb the losses and compensate for the public harm caused through the loss of natural floodplain values.

Buildings that have suffered substantial damages from flooding, fire, or other causes are not eligible for subsidized insurance. Through mortgage institutions, flood insurance is required on all buildings for which new mortgages are to be made following the publication of Flood Insurance Rate Maps.

Relief and rehabilitation efforts can be coordinated with a flood insurance scheme. The primary purpose of such coordination is to reduce flood relief costs by requiring floodplain occupants to cover their potential losses through insurance.

The National Flood Insurance Program in the United States provides both incentives and minimum requirements for local governments to manage their floodplains using nonstructural methods. The program was established by Congress under the National Flood Insurance Act of 1968. It is administered by the Federal Insurance Administration (FIA), by authority of that act, as amended, as well as the Flood Disaster Protection Act of 1973, as amended.

The Process
The National Flood Insurance Program provides flood hazard data to local governments at no cost. It also makes available subsidized flood insurance to homeowners and businesses. In return, communities must adopt and administer local floodplain management measures aimed at protecting lives and new construction from future flooding.

The Role of the FIA
The Flood Disaster Protection Act of 1973 requires the U.S. Department of Housing and Urban Development (HUD) to publish information on known flood-prone communities and to notify them of their tentative identification as such, following which the community must either make prompt application for participation in the National Flood Insurance Program within 1 year from the date FIA notifies them that they are flood-prone, or be denied federally related financial assistance.

The Role of Local Government

The FIA works closely with state and local governments and provides some technical assistance. The FIA requires local governments to adopt and enforce land use and control measures based on the technical information furnished by the FIA. These measures are:

To constrict the development of land which is exposed to flood damage where appropriate

To guide the development of proposed construction away from locations which are threatened by flood hazards

To assist in reducing damage caused by floods

To otherwise improve the long-range land management and use of flood-prone areas

The Role of Private Insurance Companies

The program enlists the support and resources of private insurance companies and financial institutions to administer the management of the floodplain. This is achieved by having insurance policies sold through private agencies in each community. Any time that a bank or savings and loan association issues a new mortgage on a floodplain building in a participating community or where federal monies are involved, a flood insurance policy is mandatory. This is enforced through the federal agencies dealing with the financial institutions.

The Role of Property Owners

As a special incentive to enter the National Flood Insurance Program, individuals and businesses located in identified areas of special flood hazard in participating communities are required to purchase flood insurance as a prerequisite for receiving any type of federally insured or federally regulated financial assistance for acquisition or construction purposes. Moreover, federal mortgage guarantees and insurance, mortgage loans, grants, and other forms of federal assistance for financing capital costs of construction and equipment are not available to individuals and businesses in identified flood-prone areas unless the community has qualified for the program by adopting effective land use and land management controls.

The Goal of the Program

The goal of the program is to limit floodplain occupancy for the short term and reduce it over a longer period. Another purpose is to reduce flood relief costs by requiring floodplain occupants to cover their own potential losses through insurance.

The Types of Perils Covered

The federally subsidized policy covers losses resulting from the perils of flood as defined in the 1968 act, subject to a $200 deductible.

The term *flood* means a general and temporary condition which may include inundation from rising waters or from the overflow of streams, rivers, or other bodies of water, or from tidal surges, abnormally high tidal water, tidal waves, tsunamis, hurricanes, or other severe storms or deluge. The policy does not cover water or mudslide damage which results solely from causes at the location of the insured property or within the control of the insured. It does not cover normal erosion losses, losses for which claims are already in progress at the time of application for coverage, or losses caused by land slippage rather than mudflow. With certain exceptions, seepage and sewer backup losses are not covered unless a general and temporary condition of the flooding exists.

Abnormal erosion caused by high-water levels accompanied by violent wave action along a lake or other body of water is considered flood damage. However, there is no coverage where normal, continuous wave action accompanied by ordinary erosion or the gradual and anticipated wearing away of the land is the proximate cause of structural property damage.

Federally subsidized flood policies have a deductible, which is applied separately to structures and to their contents.

The 1968 act provides for two specific programs: (1) the Emergency Program; and (2) the Regular Program.

The Emergency Program

A community can become eligible for flood insurance under the Emergency Program by submitting a completed application to the Federal Insurance Administrator and adopting preliminary land use measures pursuant to regulations published by FIA under the National Flood Insurance Program.

The Emergency Program is intended primarily as an interim program to provide a first-layer amount of insurance at federally subsidized rates pending publication of the FIA Flood Insurance Rate Map (FIRM).

The Regular Program

After the FIA has conducted a flood insurance study—at no cost to the community—to develop technical information, including minimum first-floor elevations for land use purposes, the data are given to the community. In addition, HUD publishes the proposed elevation levels in the *Federal Register* and twice during a 10-day period in a prominent local newspaper. The community officials and citizens are given 90 days in which to appeal the proposed elevation determinations by submitting technical data that negate or contradict the FIA's findings. During this period, flood insurance continues to be available at subsidized rates under the Emergency Program.

When the Federal Insurance Administrator makes a final determination on flood elevations, the FIA publishes a Flood Insurance Rate Map (FIRM) for determining actuarial rates. As of the date of publication of the initial FIRM, the community is converted to the Regular Program, under which additional or second-layer limits of insurance are also available at actuarial rates.

Under the Regular Program, flood insurance at first-layer limits continues to be available at subsidized rates—if less than actuarial rates—on structures existing in the community either on December 31, 1974 or prior to the effective date of the initial FIRM, whichever is later. If any new construction were to be located within identified areas of special flood hazards built subsequent to the above applicable date, it will be charged actuarial rates.

Thus flood insurance for any insurable structure or contents therein in a community in the Regular Program is available at subsidized or actuarial rates, or both.

Suspension from the Program

Participating communities may be suspended from the National Flood Insurance Program for failure to adopt land use regulations, for revoking land use regulations once adopted and approved by the FIA, or for nonenforcement of adopted land use regulations. The community's eligibility remains terminated until, upon satisfactory evidence of an active floodplain management program, the community is reinstated by the Administrator. Flood insurance cannot be

Table 8.1 Maximum Amounts of Insurance

	First layer[a] Max. amount at subsidized or actuarial rate[c]		Second layer[b] Max. additional amount at actuarial rate	
	Bldg. (each)	Contents (per unit)	Bldg. (each)	Contents (per unit)
Single-family dwelling All states and jurisdictions (except below)	$ 35,000	$ 10,000	$ 35,000	$ 10,000
Alaska, Hawaii, Guam, and Virgin Islands	50,000	10,000	50,000	10,000
Other residential (except single-family) All states and jurisdictions (except below)	100,000	10,000	100,000	10,000
Alaska, Hawaii, Guam, and Virgin Islands	150,000	10,000	150,000	10,000
Any other structure	100,000	100,000	100,000	100,000

[a]Maximim insurance available under the Emergency Program for structures in existence on and not substantially improved after December 31, 1974, or be fore the effective date of FIRM (Flood Insurance Rate Map), whichever is later. Use subsidized rates.

[b]Second-layer insurance is available under the Regular Program only.

[c]Under the Regular Program, for existing structures, use subsidized rate or actuarial rate, whichever is lower. For structures located in an identified special flood hazard area which are newly constructed or substantially im proved after December 31, 1974, or the effective date of FIRM, whichever is later, use actuarial rates only.

Source: Ref. 5.

sold or renewed in a suspended community. Policies sold or renewed in a community during a period of ineligibility shall be deemed void and unenforce able whether or not the parties to the sale or renewal had actual notice of the ineligibility.

Limits of Coverage
For existing buildings in a flood hazard area subsidized insurance is made available to the owner up to the limits given in Table 8.1. The second layer is available at actuarial rates with a limit imposed on the rate which often represents a second subsidy.

FIA Rate Tables
The FIA rate tables relate the type of building hazard exposure to the flood hazard zone through the use of various tables. Examples are given in Tables 8.2 through 8.5.

The symbols used to designate the Actuarial Rate Zones are as follows:

Zone symbol	Category
A	Area of special flood hazards and without base flood elevations determined
A1-A30	Area of special flood hazards with base flood elevations–zones assigned according to flood hazard factors
AO	Area of special flood hazards that have shallow flood depths (less than 18 in.) and/or unpredictable flow paths–base flood elevations not determined
V1-V30	Areas of special flood hazards, with velocity, that are inundated by tidal floods–zones assigned according to flood hazard factors
B	Area of moderate flood hazards
C	Area of minimal flood hazards
D	Area of undetermined, but possible, flood hazards
M	Area of special mudslide hazards
N	Area of moderate mudslide hazards
C	Area of minimal mudslide hazards
P	Area of undetermined, but possible, mudslide hazards

Subsidized rates are the rates established by the FIA which involve a high degree of participation by the federal government to encourage the purchase of first-layer limits of flood insurance at an affordable cost on existing structures.

Actuarial rates are established by the FIA pursuant to individual community flood level studies and investigations which are undertaken to provide flood insurance in accordance with accepted actuarial principles, including provisions for operating costs and allowances.

Table 8.2 FIA Subsidized Rates (Rates per $100 Insurance)

	Rate	
	Structure	Contents
A. Single-family dwelling	0.25	0.35
B. Other residential (except single-family)	0.25	0.35
C. Any other structures	0.40	0.75

Note: Under the Emergency Program, subsidized rates apply to all types of structures and contents which were in existence or on which construction or substantial improvement was started on or before December 31, 1974, or prior to the effective date of the Flood Insurance Rate Map (FIRM), whichever is later. Subsidized rates also continue to be available for all structures and contents outside the special flood or mudslide hazard area Zones A, AO, A1-A30, V1-V30, or M, regardless of date of construction.
Source: Ref. 5.

Table 8.3 FIA Zone Rate Table (Applicable Only for Communities in the Regular Program) (Rates per $100 Insurance

Zone rates apply to all buildings in a community in the Regular Program, except those buildings located in Zones A1-A30, and V1-V30. A building which was constructed or substantially improved on or before December 31, 1974, or before the effective date of the FIRM, whichever is later, may be insured for first layer amounts of insurance using Subsidized Rate Table 8-2. if **lower** than the **following** Zone Rates.

Section A—Structure—One- to Four-Family Residential

Type of Structure	A	AO	Zone B	C	D
One story—no basement	.35	.30[a]	.03	.01	.20
Two or more stories—no basement	.30	.25[a]	.02	.01	.15
Split level—no basement	.30	.25[a]	.02	.01	.15
One story—with basement	2.05	2.00	.15	.10	1.10
Two or more stories—with basement	1.30	1.35	.10	.10	.70
Split level—with basement	1.30	1.35	.10	.10	.70
Mobile home on foundation	1.40	.65[a]	.15	.15	.80

Note: The maximum actuarial rate payable by the insured on 1-4 family residential structures is $.50 for

(a) First layer limits of insurance on new construction, if first floor elevation is at or above the base flood elevation, or

(b) second layer limits of insurance on all structures.

Section B—All Other Structures

	A	AO	Zone B	C	D
One story—no basement	.60	.50[a]	.05	.02	.30
Two or more stories—no basement	.50	.40[a]	.04	.02	.25
Split level—no basement	.45	.40[a]	.04	.02	.35
One story—with basement	3.40	3.30	.25	.20	1.85
Two or more stories—with basement	2.15	2.25	.15	.20	1.15
Split level—with basement	2.15	3.00	.15	.20	1.25
Mobile home on foundation	2.30	1.10[a]	.30	.25	1.30

Section C—Contents—Residential

Location in Structure	A	AO	Zone B	C	D
All in basement	41.50	26.00	2.60	.20	22.00
All on 1st floor	.90	.75[a]	.10	.05	.50
All on 1st two or more floors	.60	.50[a]	.10	.05	.35
All on 1st floor and basement	5.65	4.00	.40	.10	3.00
All on 1st two or more floors and basement	5.90	3.50	.35	.10	3.10
All above 1st floor	.15	.05[a]	.01	.01	.08
All in mobile home on foundation	1.35	.55[a]	.10	.05	.75

Section D—All Other Contents

Location in Structure	A	AO	Zone B	C	D
All in basement	50.00	39.00	3.90	.30	50.00
All on 1st floor	1.35	1.10[a]	.10	.10	.75
All on 1st two or more floors	.85	.75[a]	.10	.10	.50
All on 1st floor and basement	8.50	6.00	.60	.15	4.55
All on 1st two or more floors and basement	5.90	5.25	.55	.15	3.20
All above 1st floor	.20	.08[a]	.01	.01	.10
All in mobile home on foundation	2.00	.85[a]	.10	.10	1.05

[a]For structures without basement located in Zone AO, where the first floor is 18 in. or more above the crown (highest point) of the nearest street, use Zone B rates.
Source: Ref. 5.

Table 8.4 FIA Elevation Rate Table III (Section A-1- to 4-Family Residential Structure, 1 Story)

ELEVATION OF FIRST FLOOR ABOVE OR BELOW BASE FLOOD ELEVATION	NO BASEMENT ZONES				WITH BASEMENT ZONES			
	A1 – A7	A8 – A14	A15 – A17	A18 – A30	A1 – A3	A4 – A7	A8 – A9	A10 – A30
+ 5 OR MORE	.01	.01	.01	.01	.10	.10	.10	.10
+ 4	.01	.01	.01	.01	.10	.10	.10	.10
+ 3	.01	.01	.02	.04	.10	.10	.10	.10
+ 2	.01	.02	.05	.08	.10	.10	.11	.13
+ 1	.01	.07	.10	.15	.90	.30	.24	.22
0	.12	.16	.19	.23	4.78	.84	.49	.33
– 1	.48	.31	.31	.34	13.13	2.13	.95	.49
– 2	1.59	.55	.47	.48	a	4.95	1.77	.71
– 3	a	.93	.70	.64	a	6.73	3.15	.98
– 4	a	1.48	1.00	.83	a	a	5.16	1.36
– 5	a	2.34	1.40	1.07	a	a	a	1.87
– 6	a	2.86	1.91	1.34	a	a	a	2.52
– 7	a	a	2.62	1.66	a	a	a	3.40
– 8	a	a	3.53	2.02	a	a	a	4.56
– 9	a	a	a	2.48	a	a	a	5.21
–10	a	a	a	3.03	a	a	a	a
–11 OR LOWER	a	a	a	a	a	a	a	a
ZONE RATE	.35	.55	.73	.95	7.36	2.01	1.33	1.12

aUse $25.00 rate.
Source: Ref. 5.

Table 8.5 FIA Elevation Rate Table 5 (Section D—Nonresidential Contents)

ELEVATION OF FIRST FLOOR ABOVE OR BELOW BASE FLOOD ELEVATION	ALL ON FIRST FLOOR ZONES				ALL ON FIRST FLOOR & BASEMENT ZONES			
	A1–A7	A8–A14	A15–A17	A18–A30	A1–A3	A4–A7	A8–A9	A10–A30
+ 5 OR MORE	.20	.20	.20	.20	.20	.20	.20	.20
+ 4	.20	.20	.20	.20	.20	.20	.20	.20
+ 3	.20	.20	.20	.20	.20	.20	.20	.24
+ 2	.20	.20	.20	.30	1.62	.53	.44	.42
+ 1	.20	.20	.38	.50	8.46	1.52	.92	.72
0	.44	.57	.65	.74	23.00	3.87	1.80	1.14
– 1	1.80	1.08	1.01	1.08	36.12	9.11	3.42	1.76
– 2	5.88	1.85	1.52	1.49	a	15.45	6.14	2.61
– 3	a	3.09	2.21	1.98	a	18.56	10.17	3.81
– 4	a	4.85	3.15	2.55	a	a	14.19	5.48
– 5	a	7.50	4.40	3.26	a	a	a	7.78
– 6	a	9.27	6.00	4.10	a	a	a	10.80
– 7	a	a	8.22	5.07	a	a	a	13.80
– 8	a	a	11.09	6.17	a	a	a	a
– 9	a	a	a	7.55	a	a	a	a
–10	a	a	a	9.23	a	a	a	a
–11 OR LOWER	a	a	a	a	a	a	a	a
ZONE RATE	1.31	1.88	2.33	2.90	27.06	8.48	4.64	3.35

aUse $50.00 rate.
Source: Ref. 5.

Payable rates are those rates shown in the Application and Declarations Form used to calculate the premium payable by the insured. These may be either subsidized or actuarial rates.

Substantial improvement means any repair, reconstruction, or improvement of the structure, the cost of which equals or exceeds 50 percent of the market value of the structure under either of the following conditions:

1. Before the improvement is started
2. Before the damage occurred (if the structure has been damaged and is being restored)

Substantial improvement is started when the first alteration of any wall, ceiling, floor, or other structural part of the building commences.

8.6 PROGRAM ELEMENTS

A nonstructural floodplain management program includes a range of measures for planning and action which must be integrated, implemented, and regularly evaluated. It can be most effective when multiple uses are included as part of the program. There are many competing demands for urban land, and, therefore, single-purpose utilization of floodplain lands for conveyance and storage of floodwaters does not usually represent optimized use.

Further, multiple use of floodplain lands also provides the basis for multisource funding and better support from the general public and private interests.

Program Integration

The nonstructural floodplain management program must represent an integration of the many components. Integration must be aimed at developing synergistic benefits.

Program Implementation

The implementation of a nonstructural floodplain management program for a drainage basin requires action at the various levels of government. Implementation would include:

    Public input to the program
    Involvement of municipal engineers and councils
    Adoption of the program by municipal councils
    Public informational effort
    Programmed surveillance to insure compliance

Program Evaluation

The nonstructural management program requires regular evaluation to measure success or failure. Benchmark conditions should be established to allow objective evaluation against a predetermined standard. Evaluations annually are appropriate.

REFERENCES

1. *Human Adjustment to Floods: A Geographical Approach to the Flood Problem in the United States*, Gilbert F. White, University of Chicago, Department of Geography, Research Paper No. 29, Chicago, 1945.
2. *A Unified National Program for Managing Flood Losses*, House Document 465, 89th Congress, 2nd Session, U.S. Government Printing Office, Washington, D.C., 1966.
3. *Flood Proofing: An Element in a Flood Damage Reduction Program*, John R. Sheaffer, University of Chicago, Department of Geography, Research Paper No. 65, Chicago, 1960.
4. *Introduction to Flood Proofing*, John R. Sheaffer, Center of Urban Studies, University of Chicago, 1967.
5. *Flood-Proofing Regulations*, U.S. Army Corps of Engineers, Office of the Chief of Engineers, Washington, D.C., June 1972.
6. *Economic Feasibility of Flood Proofing: Analysis of a Small Commercial Building*, Federal Emergency Management Agency, U.S. Government Printing Office, Washington, D.C., 1979.

9

Floodplain Regulations

9.1 INTRODUCTION

For decades developments have been encroaching upon the nation's flood-plains, ignoring their natural function of carrying floodwaters. As a result, floods have regularly taken their toll in lives and property. Over the years massive structural "solutions" such as dams and channels were built. Although $10 billion have been spent on structural works since 1936, flood losses and federal disaster relief outlays have continued to escalate. Beginning in the mid-1950s, there has been a growing realization that instead of attempting to keep floodwaters away from people, it might be more effective to keep people away from flood hazard areas. This requires floodplain delineation and regulation. This concept was accelerated when the U.S. Congress passed the Housing and Urban Development Acts of 1968 (PL 90-448) and 1969 (PL 91.152) and the Flood Disaster Protection Act of 1973 (PL 93-234).

The Housing and Urban Development Act of 1968 established a National Flood Insurance Program. An integral part of the program was the require-ment that a community wishing to have its citizens benefit from federally subsidized flood insurance would have to regulate its floodplains to mitigate future flood losses. This was necessary if the insurance program was to assist efforts to achieve the national goal of reducing average annual flood losses.

The Federal Insurance Administrator has legislative responsibilities to "develop comprehensive criteria designed to encourage, where necessary, the adoption of adequate state and local measures which, to the maximum extent feasible, will

1. constrict the development of land which is exposed to flood damage where appropriate,
2. guide the development of proposed construction away from locations which are threatened by flood hazards,
3. assist in reducing damage caused by floods, and
4. otherwise improve the long-range land management and use of flood-prone areas, and she (he) shall work closely with and provide any necessary technical assistance to State, interstate, and local governmental agencies, to encourage the application of such criteria and the adoption and enforcement of such measures." [Section 1361 (PL91:152)]

There are many different techniques by which a governmental entity can avoid development of flood hazard areas. It can purchase such lands in fee. It can purchase a flood or open-space easement which would leave the land in private ownership, but limit it to uses compatible with its function as a floodplain. The same can be accomplished by purchase of development rights.

If only a portion of the property to be developed lies in a flood hazard area, that portion can be kept open by a transfer of the overall allowed density to the upland portion. For example, if the overall density permitted in a proposed subdivision is 2 units/acre, and one-half the property is in the floodplain, then that half can be kept open by permitting 4 units/acre on the upland property. The Planned Unit Development is a similar technique when structures are clustered together, leaving expanses of open space.

Some communities require dedication of a certain percentage of land to the public in their subdivision regulations. The flood-prone portion can become the public land dedication.

Of course, these techniques may only be partial solutions if such density transfers or public land dedications would still result in unwise use of the floodplain, but they indicate some of the tools in the arsenal of floodplain management.

This chapter focuses on a major tool for floodplain management—floodplain regulations, also known as floodplain zoning. There are two basic parts. First, an outline is presented to provide the scope and contents of floodplain regulations. Second, a discussion is presented on information sources to be used in identifying areas which are to be regulated.

In adopting such regulations, a state or local government is using its police power, which is limited by constitutional and statutory provisions. The extent and limits of the power are the subject of many texts and law review articles and will not be addressed here. A number of cases on floodplain management are cited in Chap. 2. The basic contents of such regulations, however, are discussed in Sec. 9.2.

The National Flood Insurance Program has been the major impetus for the adoption of floodplain regulations by communities across the nation. The Federal Insurance Administrator (also an Assistant Administrator in the Federal Emergency Management Agency) has adopted minimum floodplain regulations which must be adopted by communities in order to participate in the National Flood Insurance Program. The Administrator, however, encourages communities to adopt more comprehensive regulations. When a community does so, the more restrictive regulations apply. Therefore, the federal requirements and criteria should be considered a "floor" but by no means the optimum criteria for any particular community.

The goal of a community's floodplain regulations should be to achieve the wise use of flood hazard areas and thereby minimize flood losses. Additional public harm is avoided by preserving natural floodplain values. Floodplains in their natural or relatively undisturbed states serve water resource values (natural moderation of floods, water quality maintenance, and groundwater recharge), living resource values (fish, wildlife, and plant resources), cultural resource values (open space, natural beauty, scientific study, outdoor education, and recreation), and cultivated resource values (agriculture, aquaculture, and forestry).

9.2 CONTENTS OF FLOODPLAIN REGULATIONS

Authority

There is often no need to state in the regulations themselves the authority by which the entity promulgating the regulation is doing so. However, the entity must have such authority for the regulations to be valid. Home rule cities may derive their authority from both the state constitutional provisions establishing municipal home rule powers and from statutes. Statutory cities derive their power from state statutes. Most states have adopted legislation which delegates general zoning powers to local governments. Such statutes authorize local governments to promulgate regulations designed to secure safety from fire, panic, and other dangers; to promote health and general welfare; and to consider the most appropriate use of land. Subdivision and building codes enabling acts also authorize regulations to protect public health and safety.

Some states have specifically added flooding provisions to such legislation. However, even where such specific wording has not been added, courts have invariably found the general enabling statutes sufficient authority for adoption of floodplain regulations.

Purposes

The purposes clause showing the intent to prevent public harm rather than to secure public benefit is a very important section. It puts the rest of the regulations into context and establishes the framework by which the other clauses are interpreted. In addition, it is crucial from a constitutional point of view when the application of the regulation to private property is tested in courts. The cases cited in Chap. 2 should be used to establish as solid a constitutional base as possible.

Definitions

This section establishes the meaning of the terms generally used in floodplain regulations. Only those terms which are actually used in the text should be defined.

Base flood: That flood which has a 1 percent chance of occurrence in any given year (also known as a 100-year flood). This term is used in the National Flood Insurance Program to indicate the minimum level of flooding to be used by a community in its floodplain management regulations.

Base floodplain: The 100-year floodplain (1 percent chance floodplain). Also see the definition of *floodplain*.

Channel: A natural or artificial watercourse of perceptible extent, with a definite bed and banks to confine and conduct continuously or periodically flowing water.

Facility: Any man-made or man-placed item other than a structure.

Flood or flooding: A general and temporary condition of partial or complete inundation of normally dry land areas from the overflow of inland and/or tidal waters, and/or the unusual and rapid accumulation or runoff of surface waters from any source.

Flood fringe: That portion of the floodplain outside of the regulatory floodway (often referred to as "floodway fringe").

Floodplain: The lowland and relatively flat areas adjoining inland and
coastal waters including flood-prone areas of offshore islands, including, at a minimum, that area subject to a 1 percent or greater
chance of flooding in any given year. The base floodplain shall be
used to designate the 100-year floodplain (1 percent chance floodplain). The critical action floodplain is defined as the 500-year floodplain (0.2 percent chance floodplain).

Flood proofing: The modification of individual structures and facilities,
their sites, and their contents to protect against structural failure,
to keep water out, or to reduce effects of water entry.

Flood profile: A graph showing the elevations of the floodwater surface
and the elevations of the underlying land as a function of distance
along a path of flows.

Flood protection elevation: An elevation 2 ft above the elevation of the
water surface of a 100-year flood.

Minimizing flood losses: Achieving an absolute decline in average annual
flood losses.

New construction: Structures for which the start of construction
occurred on or after a specified date.

One percent chance flood: The flood having 1 chance in 100 of being
exceeded in any 1-year period (a large flood). The likelihood of
exceeding this magnitude increases in a time period longer than 1
year. For example, there are 2 chances in 3 of a larger flood exceeding the 1 percent chance flood in a 100-year period.

Preserve: To prevent modification of the natural floodplain environment
or to maintain it as closely as possible in its natural state.

Regulatory floodway: The area regulated by federal, state or local
requirements; the channel of a river or other watercourse and the
adjacent land areas that must be reserved in an open manner, i.e.,
unconfined or unobstructed either horizontally or vertically, to
provide for the discharge of the base flood so the cumulative increase
in water surface elevation is no more than a designated amount (not
to exceed 1 ft, as set by the National Flood Insurance Program).

Restore: To reestablish a setting or environment in which the natural
functions of the floodplain can again operate.

Structures: Walled or roofed buildings, including mobile homes and gas
or liquid storage tanks, that are primarily above ground (as set by
the National Flood Insurance Program).

Substantial improvement: Any repair, reconstruction, or improvement of
a structure the cost of which equals or exceeds 50 percent of the
market value of the structure either (1) before the improvement or
repair is started; or (2) if the structure has been damaged and is
being restored, before the damage occurred. For the purposes of
this definition, *substantial improvement* is considered to occur when
the first alteration of any wall, ceiling, floor, or other structural
part of the building commences, whether or not that alteration affects
the external dimensions of the structure. The term does not, however, include either (1) any project for improvement of a structure
to comply with existing state or local health, sanitary, or safety code
specifications which is solely necessary to assure safe living conditions; or (2) any alteration of a structure listed on the National
Register of Historic Places or a state inventory of historic places.

Watercourse: A stream, creek, pond, natural or artificial depression, slough, gulch, arroyo, reservoir, or lake in, or into, which storm runoff and floodwater flows either regularly or infrequently. This includes established drainage ways, natural as well as man-made, for carrying urban storm runoff.

Wetlands: "Those areas that are inundated by surface or groundwater with a frequency sufficient to support and under normal circumstances does or would support a prevalence of vegetative or aquatic life that requires saturated or seasonally saturated soil conditions for growth and reproduction. Wetlands generally include swamps, marshes, bogs, and similar areas such as sloughs, potholes, wet meadows, river overflows, mud flats, and natural ponds" (Executive Order 11990, Protection of Wetlands).

Flood Regulatory Areas

This section establishes the actual areas by describing the official floodplain maps (see Sec. 9.3 for more details). These could include, for example, the Flood Insurance Rate Maps, the Flood Hazard Boundary Maps, Hydrologic Atlases, Corps of Engineers Maps, and local maps delineating the community's floodway. It also states how amendments to these official maps can be made, where the maps and changes are recorded, and how mapping disputes are to be settled.

Administration

This section states who is responsible for administering the regulations (for example, the director of public works) and describes his or her duties.

Nonconforming Structures and Uses

This section addresses existing structures and uses which were lawful prior to the adoption of the regulations but which do not conform to these regulations. Such structures and uses are "nonconforming" and may be continued, subject to certain conditions regarding expansion, improvements, and discontinuance. These conditions must comply with the minimum requirements of the National Flood Insurance Program. The intent is to avoid permitting substantial expansion or improvements which would be counterproductive to the goal of minimizing flood losses and to achieving the wise use of floodplains.

Permitted Uses in the Regulatory Floodway

The uses permitted in the regulatory floodway should not be limited to those which will not impede or affect flow or affect adversely the natural and beneficial values served by floodplains, but should be as numerous as possible to aid in defending the regulations from a successful constitutional attack under the takings clause. Other use may be permitted, but only under very stringent criteria.

Water and the adjacent floodplain exist in nature in a state of dynamic equilibrium. If one part of a coastal or riverine system is disturbed, the entire system usually readjusts toward a new equilibrium. Thus, floodplains must be carefully evaluated to identify the natural and beneficial floodplain values.

The cases cited in Chap. 2 suggest various use possibilities. The following uses are allowed within the regulatory floodway to the extent that they are not prohibited by an underlying zoning ordinance and provided they do not require any structures, facilities, fill, storage of materials or equipment, or change in a channel of a watercourse:

1. Agricultural uses such as general farming, grazing of horses and livestock, truck farming, forestry, sod farming, wild crop harvesting and the raising of plants, flowers and nursery stocks;
2. Residential uses such as lawns, gardens, driveways, and play areas;
3. Industrial-commercial types of use such as loading acres, railroad rights-of-way (not including freightyards or switching, storage or industrial sidings), airport landing strips;
4. Recreational uses such as swimming pools, golf courses, golf driving ranges, open air theaters, parks, picnic grounds, camp sites, horseback riding and hiking area;
5. Wildlife and nature preserves, game farms, and fish hatcheries;
6. Open pit mining for the removal of topsoil, sand, gravel, or other minerals;
7. Utility transmission lines, pipelines, water monitoring devices, and roadways (not including bridges).

Permitted Uses in the Flood Fringe

The flood fringe has the potential to provide natural and beneficial values to a community. These values need to be identified and evaluated before the placement of structures and facilities is considered. In addition, any plans for development must meet the goal of minimizing average annual flood losses. Within this context, new structures and facilities should be flood-proofed to or above the flood protection elevation. With respect to coastal areas, the federal regulations also require that wave heights be added to the base flood to establish design elevations.

Appeals

If an administrator will be making determinations and decisions, there should be an appeal available to the elective officials. Both a denial of a permit and the granting of a permit should be appealable. A variance bond is inappropriate for this purpose since the decisions have an enormous long-term impact.

9.3 INFORMATION SOURCES

Chapter 7 presents a discussion of methods to delineate floodplains. It is necessary to be cognizant of the availability of information on flood hazard areas from a variety of sources.

Maps showing flood hazard areas will usually be available from the National Flood Insurance Program administered by the Federal Insurance Administration (FIA). Detailed maps showing the elevations and boundaries of the 100-year (Zones A and V) and 500-year (Zone B) floodplains are known as Flood Insurance Rate Maps (FIRMs). Such maps have been published by the FIA for over 1300 communities, and maps for more communities continue to be published for the FIA's program to provide maps of all flood-prone areas by 1983. Many of the communities which have a FIRM also have a Flood Insurance Study (FIS) report containing detailed flood information. Some 13,000 less detailed maps showing the approximate areas of the base (Zone A) floodplain are available for most of the remaining communities. These are called Flood Hazard Boundary Maps (FHBMs). Similar information, some very detailed, is also available from other agencies. The search for flood hazard information should follow the sequence below:

1. The detailed map (FIRM) or the Flood Insurance Study (FIS) report should be consulted first.

2. If a detailed map (FIRM) is not available, obtain an approximate boundary map (FHBM) from the same source as in the preceding step. If the proposed site is at or near the 100-year boundary, if data on flood elevations are needed, or if the map does not delineate the flood hazard boundaries in the vicinity of the proposed site, seek detailed information and assistance from the agencies listed in Table 9.1.

Table 9.1 Sources of Floodplain Information and Technical Assistance Services for Determining Whether a Location is in a Floodplain

Agency	Floodplain maps and profiles		Technical assistance services
	Riverine	Coastal	
Department of Agriculture: Soil Conservation Service	X	X	X
Department of the Army: Corps of Engineers	X	X	X
Department of Commerce: National Oceanic and Atmospheric Administration	...	X	X
Department of Housing and Urban Development:			
Federal Housing Administration	...	...	X
Federal Insurance Administration	X	X	X
Department of the Interior:			
Geological Survey	X	X	X
Bureau of Land Management	X	X	...
Bureau of Reclamation	X	...	X
Tennessee Valley Authority	X	...	X
Delaware River Basin Commission	X	X	X
Susquehanna River Basin Commission	X	...	X
States	Varies from state to state		

3. If an approximate boundary map (FHBM) is not available or if the map does not delineate the flood hazard boundaries in the vicinity of the proposed site, seek detailed information and assistance from the agencies listed in Table 9.1.

4. If the agencies listed do not have or know of detailed information and are unable to assist in determining whether or not the proposed site is in the base floodplain, seek the services of a licensed consulting engineer experienced in this type of work. The quality of information obtained from the consulting engineer must be comparable to that required of flood insurance study contractors for the FIA (see Chap. 7).

9.4 EXAMPLE OF A COMPREHENSIVE ORDINANCE

The DuPage County Floodplain Ordinance is presented as an example that addresses both flood loss reduction and the preservation and the enhancement of natural floodplain values. This ordinance has been administered successfully for more than 5 years.

Rules and definitions of the DuPage County Zoning Ordinance, to be amended to provide for the following new definitions:

Floodplain: The land area adjacent to surface water bodies or waterways that is subject to periodic inundation when greater than normal water flows are experienced. The floodplain may be identified according to the frequency of the flood flow that inundates it. For example, the portion of floodplain inundated by a hundred year storm is the 100-year floodplain. The special flood hazard area is the floodplain within a community subject to inundation by the 100-year flood—the flood having a one percent chance of being equalled or exceeded in any given year.

Wetlands

1. The land areas in which the groundwater table or zone of saturation periodically intersects the surface and which contain wetland plant species that are listed in ...*A Manual of Aquatic Plants* [9]...and/or... *Plants of the Chicago Region* [10]...and/or...*The New Britton and Brown Illustrated Flora of the Northeastern United States and Adjacent Canada* [11].... Wetland plants are distinguished from aquatic plants in that they are limited to species that can complete their life cycle without submersion but which have the ability to withstand a permanent or seasonally long submersion of at least the plant root system.

2. Wetlands, for the purposes of this ordinance, are also defined to include areas of poorly drained soils with the water table within two (2) feet of the ground surface for at least three (3) months of the year.

Minimum floodplain elevation: That elevation determined from the flood crest profile of the highest flood of record. The flood crest profiles are a part of the Hydrologic Investigation Atlases. Precise elevation determinations shall be made by drawing a line from the said property to the centerline of the channel and perpendicular to said centerline to determine the proper station, and then reading the proper elevation of the flood crest on the flood crest profile. Detailed flood crest profiles are kept on record in the offices of the DuPage County Department of Public Works, the DuPage County Planning and Zoning Departments and the

Northeastern Illinois Planning Commission. In case of a dispute over the interpretations of the map elevations, the decision of the Superintendent of DuPage County Department of Public Works shall govern.

Intermediate Regional Flood: A flood having an average frequency of occurrence on the order of once in 100 years (i.e., with a 1% chance of being equalled or exceeded in any given year). This flood is based on a statistical analysis of streamflow records available for the watershed, and analysis of rainfall and runoff characteristics in the general region of the watershed. Such a flood may, of course, occur in any year or at any time, although its statistical frequency is once in 100 years. Detailed maps are on file in the offices of the DuPage County Department of Public Works, the DuPage County Regional Planning Commission, and the Northeastern Illinois Planning Commission.

Standard Project Flood: The flood that may be expected from the most severe combination of meteorological and hydrological conditions that are considered reasonably characteristic of the geographical area in which the drainage basin is located, excluding extremely rare combinations. As used by the U.S. Army Corps of Engineers, Standard Project Floods are intended as practicable expressions of the degree of protection that should be sought in the design of flood control works, the failure of which might be disastrous.

Flood crest: The maximum stage or elevation reached by the waters of a flood at a given location....

Floodplain

This section of the DuPage County Zoning Ordinance is designed to balance future development and public and private interests with physical environmental conditions by regulating the use of all floodplains and/or wetland areas in DuPage County, so as to prevent the public harms associated with flooding, loss of natural floodwater storage, loss of groundwater supply, and disturbance of natural ecosystems.

1. Methodology of Minimum and Intermediate Regional (100-Year) Floodplain Delineation

The following information, consistent with accepted engineering standards, shall be used to delineate the Minimum Floodplain and Intermediate Regional (100-year flood) Floodplain area(s) of DuPage County.

a. Hydrologic Investigation Atlas, Series IIA, as amended from time to time as follows:

HA-68 Elmhurst Quadrangle
HA-70 Aurora North Quadrangle
HA-86 Hinsdale Quadrangle
HA-142 Geneva Quadrangle
HA-143 Lombard Quadrangle
HA-146 Romeoville Quadrangle
HA-148 Wheaton Quadrangle
HA-149 Sag Bridge Quadrangle
HA-154 Naperville Quadrangle
HA-202 West Chicago Quadrangle
HA-210 Normantown Quadrangle
Aurora South Quadrangle

b. Geologic Environments map prepared by the Illinois State Geological Survey (1968);

c. Soil Identification Atlas Sheets prepared by the United States Department of Agriculture, Soil Conservation Service;

d. Map of Piezometric Surface of Shallow Dolomite prepared by the Illinois State Water Survey (Summer 1966) and *Ground-Water Resources of DuPage County, Illinois* [12]...;

e. Flood Hazard Boundary Map and Flood Insurance Rate Map which are issued by the Administrator of the Federal Insurance Administration.

The definition of the minimum floodplain elevation shall be applied at all locations on the East and West Branches of the DuPage River and tributaries for which the mapping of the intermediate regional (100-year) floodplain by the U.S. Department of the Army, Corps of Engineers is not yet complete, until such mapping is complete; and at all locations along Salt Creek and its tributaries for which the mapping of the intermediate regional (100-year) floodplain by the U.S. Department of Agriculture, Soil Conservation Service is not yet complete until such mapping is complete. In the case of a discrepancy between two or more sources of data, the most recent information shall be used. Where no published information exists for a piece of property, a survey shall be made by a registered professional engineer to determine the 100-year floodplain elevation.

2. *Purpose*

The flood hazard areas of DuPage County, Illinois are subject to periodic inundation which results in loss of life and property, health and safety hazards, disruption of commerce and governmental services, extraordinary public expenditures for flood protection and relief, and impairment of the tax base all of which adversely affect the public health, safety and general welfare. These flood losses are caused by:

1. the cumulative effect of obstructions in floodplains, causing increased flood heights and velocities and

2. the occupancy of flood hazard areas by uses vulnerable to floods or hazardous to others which are inadequately elevated or otherwise protected from flood damages.

It is the purpose of this Ordinance to promote the public health, safety and general welfare and to minimize those losses described above by provisions designed to restrict or prohibit uses which are dangerous to health, safety, or property in times of flood or which cause excessive increases in flood heights or velocities; and by provisions that require that uses vulnerable to floods, including facilities which serve such uses, be protected against flood damage at the time of initial construction.

In addition it is also the purpose of this Ordinance to prevent public harm of a nature other than simply the flood losses described above. Public harm in this context refers to the adverse impacts of unsound land use decisions and excessive disturbance of the natural environment. This ordinance, by regulating the kinds of land use activities that may take place in the floodplains and wetland areas of DuPage County, prevents the public harm that are associated with flooding, loss of natural floodwater storage areas, loss or contamination of the groundwater supply, and the disturbance of natural ecosystems.

To further the overall goals of protecting life and property from flood damages and preventing public harm, the following specific criteria are established to manage and control the use of floodplain and/or wetland areas and to evaluate requests for special use permits.

a. To consider flooding and drainage problems as space allocation or land use problems and to the degree practicable allocate the natural storage space—floodplains and wetlands—for this purpose.

b. To use floodplains and wetlands for land use purposes which are compatible with their physical conditions, thereby preventing the creation of new flood, and other related drainage problems.

c. To avoid the increase of upstream and downstream flood risks by preventing increased run-off and a redistribution of flood waters.

d. To preserve and possibly enhance the quantity of groundwater supplied by maintaining open and undisturbed prime natural groundwater recharge areas.

e. To preserve and possibly enhance the quality of ground and surface water resources by helping to control excess infiltration into sanitary sewer systems and by preventing the location of septic systems and mixed fills in unfavorable physical environments.

f. To preserve distinctive floodplain and wetland flora and fauna and their natural inhibitants continued therein.

g. To consider floodplain and wetland areas in density or land use intensity consideration in zoning and subdivision determinations.

h. To lessen or avoid the hazards to persons and damage to property resulting from the accumulation of runoff of storm or flood waters as essential for the health, safety and general welfare of the people of DuPage County, in accordance with Chapter 34, Section 3151 of the Illinois Revised Statutes.

i. To assure that all special use permits require compliance with the minimum Rules or Regulations of the National Flood Insurance Program as issued by the Federal Insurance Administration [13].

3. *General Provisions for Minimum Floodplain or Intermediate Regional Floodplain*

Except by special use ordinance no minimum floodplain or intermediate regional floodplain and/or wetland area(s) within DuPage County shall be disturbed, reshaped, or otherwise affected by: channel relocation, channel deepening, filling or grading of any type, or the erection of any structure(s) (including but not limited to buildings, culverts, docks, dams, or bulkheads), or the storage of any materials or equipment. A special use ordinance will not be required for agricultural uses or normal maintenance of waterways, provided such maintenance is limited to the removal of accumulated debris and the control of vegetation which has the potential to interfere with flows in the waterways. Materials removed must be placed in sites approved by the DuPage County Health Department and reviewed by the DuPage County Regional Planning Commission.

4. *Permitted Uses in the Minimum Floodplain or Intermediate Regional Floodplain*

When permitted by the underlying base zoning district, the following uses shall be permitted within the minimum floodplain or intermediate regional floodplain and/or wetland area, to the extent that they are not prohibited by any law or other provision of this ordinance: agricultural

uses without permanent buildings, such as, but not limited to general
farming, pasture, grazing, outdoor plant nurseries, floriculture,
apiculture, horticulture, viticultural, truck farming, forestry, sod
farming and wild crop harvesting; private or public recreational uses
without permanent buildings, such as golf courses, golf driving ranges,
archery ranges, picnic grounds, boat launching ramps, swimming areas,
parks, camping grounds, wildlife and nature perserves, game farms,
shooting preserves, target ranges, trap and skeet ranges, hunting and
fishing areas, hiking, horseback riding and bicycle riding; and portions
of industrial, commerical, institutional, and residential uses such as
natural areas, lawns, gardens, and unpaved play areas.

REFERENCES

1. *Preliminary Nonstructural Floodplain Management Plan for the DuPage
    River* (draft), Sheaffer & Roland, Inc., prepared for the U.S. Army
    Corps of Engineers, Chicago, 1978.
2. *Economic Feasibility of Floodproofing—Analysis of a Small Commercial
    Building*, Sheaffer & Roland, Inc., Federal Energy Management Agency,
    Washington, D.C., 1979.
3. *Elevating to the Wave Crest Level—A Benefit:Cost Analysis*, Sheaffer &
    Roland Inc., Federal Emergency Management Agency, Washington, D.C.,
    1980.
4. *Alternatives for Implementing Substantial Improvement Definitions*,
    Sheaffer & Roland, Inc., Federal Emergency Managment Agency, Wash-
    ington, D.C., 1980.
5. *Floodplain Management Guidelines for Implementing E.O. 11988*, U.S.
    Water Resources Council, Washington, D.C., 1978; reprint of *Federal
    Register*, Part VI, Volume 43, No. 29, February 10, 1978.
6. *A Unified National Program for Flood Plain Management*, U.S. Water
    Resources Council, Washington, D.C., 1979.
7. *Changes in Urban Occupance of Flood Plains in the United States*, Gilbert
    F. White, Wesley C. Calef, James W. Hudson, Harold M. Mayer, John R.
    Sheaffer, and Donald J. Volk, University of Chicago, Department of
    Geograph Research Paper No. 57, Chicago, 1958.
8. *Flood Hazard in the United States: A Research Assessment*, Gilbert F.
    White, University of Colorado, Institute of Behavioral Science, Program
    on Technology, Environment, and Man Monograph #NSF-RA-E-75-006,
    Boulder, 1975.
9. *A Manual of Aquatic Plants*, Norman C. Fassett, University of Wisconsin
    Press, Madison, 1957.
10. *Plants of the Chicago Region*, 2nd ed., Floyd Swink, Morton Arboretum,
    Lisle, Ill., 1974.
11. *The New Britton and Brown Illustrated Flora of the Northeastern United
    States and Adjacent Canada*, Henry A. Gleason, Hafner Press, New York,
    1952.
12. *Ground-Water Resources of DuPage County, Illinois*, Illinois State Water
    Survey and Illinois State Geologic Survey, Cooperative Ground-Water
    Report 2, Urbana, Ill., 1962.
13. *Federal Register*, Vol. 41, No. 207, Tuesday, October 6, 1976.

10

Man-Made Storage Planning Concepts

10.1 INTRODUCTION

The essence of managing runoff is the realization that storm runoff is a space allocation problem. That is, at any given place and time, there is a fixed volume of water in storage or transit for any given rainfall event.

Natural vs. Man-Made Storage

Runoff storage exists in all natural settings. This storage occurs in many forms, a few of which are:

Water held by vegetation
Water infiltrated and held in the soil and other substrata
Water held in small, shallow surface depressions
Water held in large surface depressions or ponds and lakes
Water dynamically stored in streams and floodplains

This storage is referred to as *natural storage*. Urbanization, agricultural, and forestry practices can modify the natural storage and the stream network. Compensation for these effects, or the desire to modify the drainage system, often leads to man-made storage to replace and/or augment natural storage.

An effect of urbanization is generally to increase the rate of runoff response due to faster hydraulic conditions that exist in paved areas vs. vegetated areas. An objective of storage is to slow this rate of response of the development area. By using slow-flow channels, revegetation, and planned storage, the effects of urbanization are minimized, and in many cases planned positive results can occur.

Man-Made Storage Potential

Construction of man-made storage is a valuable means to achieving well-managed runoff systems. Such storage can vary in scale from large retarding (detention) basins on principal rivers and creeks which regulate flood flows to rooftop ponding systems which would control a wide range of runoff flows to help reduce local storm drainage works.

Storage facilities can be managed to provide multiple benefits. Such benefits include water quality improvement, sediment control, water supply, and recreational opportunities. In some instances, valley configurations are conducive to the development of storage sites. Such storage can be achieved by a road fill across a valley. Excavations for aggregate or fill can also provide storage opportunities.

Man-made storage must be viewed as only one of the possible measures to be considered in a drainage program. It must be coordinated with efforts to maintain and possibly enhance the natural storage.

Man-made storage should be planned initially in terms of drainage requirements. Aesthetics and/or recreational considerations must be subordinated to that purpose. Such storage should be evaluated in regard to economic feasibility and physical practicability.

When provision of storage is being considered, the designer must verify that the attentuation of the peak runoff will not undesirably aggravate any potential downstream peaking conditions for a range of flood frequencies. Consideration must also be given to the effect of the prolongation of flows. Assessment of these aspects must not be limited to the immediate watercourse or watercourses under consideration, but must extend to any other watercourse along which the floodwaters are conveyed. In some instances this may necessitate routing runoff from storms of durations critical for each reach of the watercourses under consideration through the whole of the drainage system upstream of the reach; in other instances, only a superficial assessment based on experienced judgment may suffice.

The greater the number of storages in a system, the more complex is the analysis of the interaction of the various outflows. Also, for such storages to function in accordance with their design intent during a given event, they must be regularly and effectively maintained. This factor must be taken into account by the designer.

Location of Man-Made Storage

Man-made storage can be located throughout a catchment. To be effective, the storage must be related to the area to be protected. With respect to location, man-made storage can be located upstream of the area to be protected, (dispersed) within the areas to be protected, and downstream from the area to be protected. The location selected will be determined by the nature and source of the flood problem.

Upstream Storage

This storage takes place upstream from the area to be protected. Its purpose is to store runoff which originates upstream or beyond the area to be protected.

Within-Area Storage

This storage takes place within the area to be protected. It can be dispersed throughout the area. Its primary purpose is to provide storage for the increased runoff which results from the urban development. Frequently such storage is provided at the development sites.

Downstream Storage

This storage takes place downstream from the area to be protected. Its purpose is to provide acceptable outlets for discharges from a storm sewerage system. In general, downstream storage manages the runoff from the protected area and mitigates any downstream effects that may be associated with development in the protected area.

Types of Storage

Storage facilities can be classified into three basic types. These types provide a range of managment opportunities to the designer. Within a given catchment, a plan may use a mix of the three types of storage.

Retention Storage

Retention storage is provided in a basin in which the runoff from a given flood event is stored and is not discharged into the downstream drainage system during the flood event. This type of stored water may be used for beneficial purposes such as irrigation or low-flow augmentation or be allowed to evaporate or seep into the ground. To be totally effective, the stored water in the flood control part of the basin must be used or lost before the next flood event occurs. A permanent conservation pool can be designed into a retention storage facility. When this is done, the facility may be referred to as wet storage.

Retardation (Detention) Storage

Retardation storage is short-term storage which attenuates the peak flow by reducing the peak outflow to a rate less than the peak inflow and thereby lengthens the time base of the hydrograph. The total volume of water discharged is the same; it is simply distributed over a longer duration. The retarding basin usually drains completely in less than a day. The area is normally dry and can be used for recreational purposes. On rare occasions, the storage of runoff may conflict with the planned recreational use of the site.

Conveyance Storage

During the period that channels, floodplains, drains, and storm sewers are filling with runoff, the waters are being stored in a transient form. This type of storage is known as conveyance storage. Construction of slow-velocity channels with large cross-sectional areas assists in the accomplishment of such storage. In a 25-m wide channel, a 1-m increase in water level will provide approximately 25,000 m^3 of storage in a 1-km reach.

Inflow-Outflow Options

Irrespective of the location or type of storage, a designer has options with respect to the management of the inflow and outflow of the runoff.

Gravity Inflow

When the elevation of the storage area is at or below the elevation of the water to be stored, the runoff will flow into the storage area by gravity. When the storage takes place in the valley, an embankment across the channel or an excavation in the valley floor will provide for the man-made storage. The inflow into such facilities will be by gravity.

Pumped Inflow

When the storage area is at an elevation above the elevation of the water to be stored, the inflow must be pumped. Such a facility would be designed to take advantage of a prime storage site.

Gravity Outflow

When the receiving stream is at an elevation far below the elevation of the stored water, the outflow release will be by gravity. Such release is typical of all types of storage created by embankments across a channel.

Pumped Outflow

Storage in excavations is frequently below the level of the receiving stream.
In such situations, the outflow must be pumped.

General Requirements

All man-made storages should be planned to meet general requirements to
provide safe facilities that will help achieve the following goals and objectives.

> Facilities should be coordinated with the development goals and objec-
> tives and the existing land use.
> Facilities should be designed to protect against failure that would in-
> crease the potential for downstream flood loss.
> Facilities should be evaluated with consideration of normal flow condi-
> tions, frequent events, less frequent intense events such as the 100-
> year frequency rainfall event, and maximum probable events. The
> evaluation of such considerations will assure that the storage does
> not worsen downstream flood conditions.
> Facilities should be designed with careful attention to a particular design
> event. A design rainfall probability of 1 percent should normally be
> used unless specific minor facilities are being evaluated.
> Facilities should be planned with respect to the topography, soil, and
> geology.
> Facilities should be planned to reduce, to the degree practicable, opera-
> tion, maintenance, and administrative needs.
> Provisions should be made to assure the maintenance of the facilities
> over their design life.
> Floodplains should be regulated downstream of new storage facilities to
> prevent new encroachment into the area protected by the storage.
> A storage facility should not encourage creation of new flood hazards
> or set the stage for larger disasters than formerly.

10.2 STORAGE CONCEPTS AND METHODS

The degree of peak flow reduction achieved through storage can vary in
accordance with design [8]. Ideally, it should prevent flow increases. Storage
reduces downstream flood flows by holding water upstream and lengthening
the time of concentration. Analyses of the rational formula or storm runoff
models demonstrate that the quicker stormwater runs off, the greater the
instantaneous flow downstream. Man-made storage will reduce such flooding,
and it is consistent with the goal to implement a comprehensive floodplain
management program.

    This section briefly explains some of the concepts and methods that can
be used to obtain retention and retardation storage. Usually, any particular
method can be adapted for a selected design event. However, the detailed
appurtenances, size of the facility, and operating characteristics should be
adapted to meet specific needs.

    There are many methods available to achieve a desired degree of storage.
The most practical are those which fit in with the building site plan or subdivision layout in the most functional, aesthetic, and economic manner.

    Stormwater runoff will be stored. The only choice is where. If it is
stored near where precipitation occurs, it will not occupy high-cost space

in storm sewers and main drains at critical flooding times. Such storage can be developed in or adjacent to high-value business areas.

Storage of runoff has the potential to produce the following benefits:

Reduced localized flooding problems
Reduced storm sewer costs because of a smaller storm sewer system in terms of size and length (less need to commence the storm sewer system as far up in the particular basin)
Improved water quality
Mitigated erosion in small tributaries because of reduced flowrates
Prolonged rainfall runoff response time
Improved opportunities for reuse of runoff and recharge of aquifers
Reduced downstream flood flows in the main drains and rivers
Mitigated increases in rates of runoff to help avoid major hydraulic channel regime changes, which then reduce the need for downstream works

Magnitude of Storage

The volume of storage required to prevent storm runoff from increasing over present conditions can be computed.

Drainage planning is generally oriented toward management of the 1 percent (100-year) flood. This section presents general planning guidelines for sizing man-made storage facilities for volume and discharge capacities. Such storages should be reviewed and evaluated against several design considerations.

The volume of storage required is based upon several factors:

Predevelopment infiltration and detention character of soil
Percent of impervious area resulting from urbanization
Area of development
Timing of rainfall runoff response

Three methods are described below for computing the amount of retarding storage required to achieve reasonable control of runoff water from urbanized areas.

Excess Precipitation Method

The determination of storage requirements using a computational technique based on excess precipitation has been developed. The Excess Precipitation Method results in an approximate volume of storage capacity for conceptual planning studies. This method assumes that the runoff from impervious areas in the new development is the key parameter causing increased flows; therefore, if the runoff volume from these areas is retarded or retained, increased runoff will be minimized. A four-step process is involved:

Step 1. Determine the depth of design rainfall excess for the predevelopment natural condition of the tract of land to be developed. The factor will be expressed in millimeters of depth of excess precipitation, or excess precipitation height (EPH).

Step 2. Determine depth of the design rainfall excess for a unit of impervious area representing postdevelopment conditions as explained above. The factor will be expressed in millimeters of depth of excess precipitation on impervious areas (EPI). It will be greater than that for natural conditions.

Step 3. Measure in hectares the total impervious area in the posturbanized development (A_{IMP}).

Step 4. Multiply the total impervious area determined in step 3 by the difference between the excess precipitation factor computed in steps 1 and 2 above. The formula is:

$$V = (\text{EPI mm} - \text{EPH mm}) \times A_{IMP} \text{ hectares} \times 10$$

where V = storage requirement, m^3.

Micro-Macro Basin Method

Computer models may be used for determining the volume of storage required. Such models may be used for storage planning and design for large developments, municipalities, and regional facilities on a comprehensive basis.

There are no strict rules in computer modeling, but the suggested approach herein is the Micro-Macro Basin Method. Simply, the concept is to analyze several typical development proposals on a detailed basis. This portion of the analysis is referred to as the Micro. Standard reservoir routing techniques described by many references are used in the Micro analysis to accurately route the given inflow hydrograph. The results, or hydrograph responses, of the Micro analysis are then scaled to other typical development subbasins in the Macro model for the entire basin. An iterative testing procedure refines the necessary storage requirements in the local development and in the basin as a whole.

Soil Conservation Service Approximation Method [1]

An approximate method has been developed by the Soil Conservation Service (SCS) for quickly analyzing effects of storage reservoirs on peak discharges. The method is based on average storage and routing effects for many structures. The storage indication method of routing was used. Figure 10.1 relates the volume of inflow to volume of required storage for a range of peak release rates. Figure 10.2 relates the peak outflow/inflow ratio to the storage/runoff volume ratio where a single-stage pipe spillway or weir is used. Emergency spillway flow is not considered.

The accuracy of the curves in Figs. 10.1 and 10.2 depends on the relationship between the storage available, the inflow volume, and the shape of the inflow hydrograph. When only a small volume is available for temporary storage, the shape of the outflow hydrograph is very sensitive to the rate of rise of the inflow hydrograph. Conversely, when a large volume is available for storage, the shape of the inflow hydrograph has little effect on the outlow hydrograph which, in this case, is controlled by the hydraulics of the structural system. Therefore, parameters such as runoff curve number and time of concentration, which affect the rate of rise of a hydrograph, become significant parameters in analyzing the effects of structures when the peak outflow rate approaches the peak inflow rate.

In Fig. 10.1 the peak inflow rate is not a factor in determining storage requirements. It can be seen that the ratio of volume of storage V_s to volume of runoff V_r is relatively high. Therefore, inflow peak is not a significant parameter. Figure 10.1 is usually accurate within 5 percent for release rates under 100 $\text{ft}^3/(\text{sec})(\text{mi}^2)$ and within 10 percent for release rates over 100 $\text{ft}^3/(\text{sec})(\text{mi}^2)$.

Figure 10.2 relates the ratio of peaks to volumes. For this case, the parameters affecting the shape of the hydrograph are important. In situations where runoff curve numbers are less than 65 in combination with

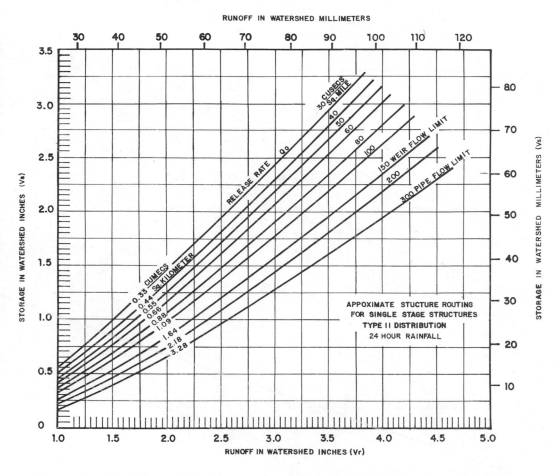

Figure 10.1 SCS graph for sizing a flood control storage structure (smaller areas): approximate single-stage structure routing for weir flow structures up to 1.64 m³/(sec)(kw²) [150 ft³/(sec)(mi²)] release rate and pipe flow structures up to 3.28 m³/(sec)(km²) [300 ft³/(sec)(mi²)] release rate. (Reproduced from Ref. 1.)

short T_c values, V_s/V_r values read from the curve will be up to 25 percent too high. Runoff curve numbers over 85 with long T_c values cause V_s/V_r values to be up to 25 percent too low.

Figure 10.1 applies to pipe drop inlets of 0 to 3.28 m³/(sec)(km²) [300 ft³/(sec)(mi²)] release rate and weir flow structures of 0 to 1.64 m³/(sec) (km²) [150 ft³/(sec)(mi²)] release rate. Figure 10.2 applies to pipe drop inlets of over 3.28 m³/(sec)(km²) [300 ft³/(sec)(mi²)] release rate and weir flow structures of over 1.64 m³/(sec)(km²) [150 ft³/(sec)(mi²)] release rate.

Extrapolation for points falling outside the limits of the curves could introduce a significant error. The steps necessary to use the procedure described in this chapter are:

1. Determine the basic watershed parameters.

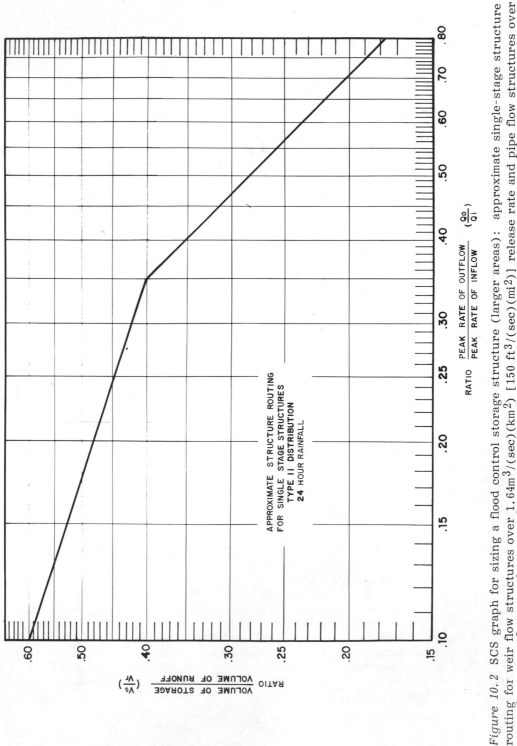

Figure 10.2 SCS graph for sizing a flood control storage structure (larger areas): approximate single-stage structure routing for weir flow structures over 1.64m³/(sec)(km²) [150 ft³/(sec)(mi²)] release rate and pipe flow structures over 3.28 m³/(sec)(km²) [300 ft³/(sec)(mi²)] release rate. (Reproduced from Ref. 1.)

2. Determine the volume of runoff and peak rate of flow from the watershed. Typically, this would be for 100-year existing or historic conditions and for future conditions.
3. Set the desired rate of outflow from the structure. This flow is normally related to historic conditions.
4. Determine the required volume of storage from the appropriate figure, Fig. 10.1 or 10.2.
5. Proportion the storage structure so that the design outflow rate and maximum storage occur at the same stage.

Note that in steps 3 and 4, the storage volume could be set and the resulting rate of outflow determined from Figs. 10.1 and 10.2. For structures with drainage areas over 8 km^2 (3 mi^2) and for events of less than 2-year frequency, other, more sophisticated techniques should be used.

Retardation Storage Release Rates

As a guideline a storage facility should be designed with a maximum release rate of 85 percent of the predevelopment flow rate for the design event unless special analysis is undertaken.

Design Criteria for Storage

The design criteria are divided into four categories in accordance with the objectives listed previously.

Control 1 Percent Runoff Rate
Use outlet works sized to release flow at 85 percent of predevelopment conditions flowrate, or less, when retardation storage is at design level.

Provide Safety against Failure
Provide safe overflows for passing storm runoff greater than the 1 percent event to avoid potential failure during very large precipitation events.

When storage is created by embankments, the embankments shall be protected against rapid washout during the maximum probable precipitation. Protection may consist of providing for wide, shallow flow over a flatly sloped embankment, an embankment core having interlocked concrete rubble or a buried retaining wall, or a designed spillway.

Minimum Operation and Maintenance
Storage should generally be incorporated into permanent landscaping, into permanent parking lot shaping or permanent features of sports ovals or parks.

Properly designed storage should be fully operable without operational care, i.e., manual or automatic valves are not desirable.

Storage should not require any unusual maintenance to insure its proper functioning when needed.

When permanent conservation pools are provided, excessive shallow areas of water should be avoided to prevent unsightly mud flats when water levels are low and to control undesired weed growth.

Consider Frequent Events
It may often be desirable to store runoff from frequent events so that the system mitigates the increased flows of these lesser events and related water quality degradation.

Pollution and Sediment Control

Chapter 11 on runoff water quality management describes water quality aspects of runoff and management techniques that can be used.

Storage provides an important tool in water quality management. Water quality and water quantity should be managed together with conservation of water as a policy keystone.

Detaining or retaining storm runoff provides an opportunity for time, sunshine, air, and gravity to assist in water quality enhancement. If special treatment is necessary, storage provides for lower rates of flow handling and therefore cost savings.

The storage treatment process can be accelerated by artificially introducing air into the storage in a manner similar to that used for aerated lagoons by sanitary engineers.

In some storage, phragmites can be planted to provide the same type of advantages given by a marshy area for water quality enhancement. Phragmites are a widely distributed seedlike grass with tall stems and large, showy, plumelike panicles.

Methods of Storage

The potential methods of developing man-made storage are presented in Table 10.1. This list of methods is illustrative of the range of choices. Storage must be site-specific. The means to be applied will reflect the proposed specific development and the site conditions.

These techniques are usually controlled by the planner in the early stages of the development. However, the architect, engineer, home builder, land developer, and government officials all have a responsibility to work toward the concept of storage which fits within the framework of a comprehensive, basinwide strategic drainage plan. Using this approach, the concepts described in the following text can effectively reduce urban costs through multipurpose use for drainage, parking, recreation, and open space, both downstream of the development and within the development.

Storage can be planned for an individual residential, commercial, or industrial parcel; an entire subdivision, office complex, or industrial park; and an entire catchment [2-7]. These storage methods are briefly discussed in the following sections.

Table 10.1 Examples of Storage Methods

Residential lots	Commercial and industrial parcels
1. Driveway storage	1. Rooftop storage
2. Cistern/infiltration	2. Parking lot storage
3. Cistern/irrigation	3. Cistern/infiltration
4. Landscaped depressions	4. Cistern/irrigation
	5. On-site ponds
Subdivisions, office complexes, industrial parks	Catchment
1. On-site ponds	1. Retention reservoirs
2. Parking lot storage	2. Dentention reservoirs
3. Slow-flow drainage patterns	3. Gravel pits and quarries
4. Open space storage	4. Open space storage
5. Rooftop storage	

Driveway Storage
Driveways can be constructed so that runoff from the lot and/or roof is routed to a depressed section of driveway. The design of the outlet system will regulate the discharge into the drainage system.

Cistern/Infiltration
Runoff from the lot and particularly from roofs can be routed to a buried cistern or tank of adequate volume and with emergency overflow. Depending upon the subsurface soils and geologic conditions, the water can be infiltrated after the storm subsides.

Cistern/Irrigation
Alternatively to the above method, the water in the cistern can be used for an irrigation water supply or discharged into the storm drain system. In areas of rolling terrain, the irrigation water may be distributed by a spreader pipe by gravity.

Rooftop Storage
Storage of water on flat commercial or industrial roofs can often be economically achieved. Roofs are usually designed to be able to support adequate loads. A special drainage outlet is provided to regulate the release of the water.

A typical rooftop storage ring is shown in Fig. 10.3. The ring is placed around the standard roof drain outlet specified in the building code. Bottom holes permit small flows to reach the roof drain unimpeded. The ring and spacing of the upper orifice in the ring are designed to allow the reduced 1 percent flow to flow unimpeded to the roof drain. Any larger volume of storage will overtop the ring unimpeded and flow normally to the roof drain. Maximum average depth of controlled ponding is usually not more than 10 cm (4 in.), representing a hydraulic load of 100 kg/m^2 (20 lb/ft^2). Structural engineers should verify roof-loading capability on existing buildings and should include roof storage loads when designing new buildings.

Parking Lot Storage
Grading of parking lots for storage is one of the least troublesome and most effective means of reducing runoff. Grading routes runoff to storage areas where controlled outlets are placed. Outlets are either grated storm inlets sized to restrain the outflow or cuts in surrounding low berms or concrete retaining walls sized to regulate the design flow.

Grading of the pavement surface should be accomplished to minimize conflict between use of the lot and storage of storm runoff. However, storage of runoff is appreciable only several times each 10 years, and even then, storage may occur during times of nonuse. Maximum depth of storage would occur only for a short time, about once each 100 years. Conflict in uses is not a significant problem.

On-Site Ponds
The construction of on-site ponds which have aesthetic and recreational benefits provides significant storm runoff detention benefits when properly planned and designed. Such ponds can be designed as common open space or incorporated as green median strips in a site development plan.

The hydraulic design of the storage function of permanent decorative ponds is based upon surcharge of the pond during precipitation events. Outflow of storm runoff is controlled by overflow weirs or orifices. This is effective ponding and can be implemented in high-density areas.

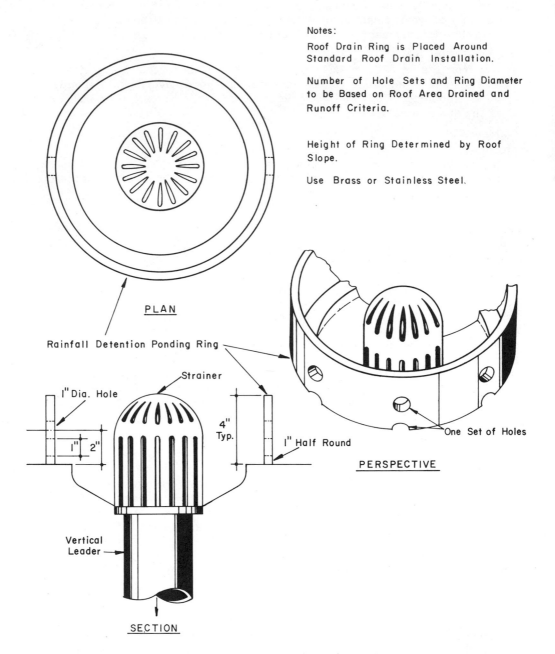

Notes:

Roof Drain Ring is Placed Around Standard Roof Drain Installation.

Number of Hole Sets and Ring Diameter to be Based on Roof Area Drained and Runoff Criteria.

Height of Ring Determined by Roof Slope.

Use Brass or Stainless Steel.

PLAN

Rainfall Detention Ponding Ring

Strainer

1" Dia. Hole

4" Typ.

1" 2"

1" Half Round

One Set of Holes

PERSPECTIVE

Vertical Leader

SECTION

Figure 10.3 Rainfall detention ponding ring for flat roofs—urban drainage, upstream ponding.

Slow-Flow Drainage Patterns

This method can be used in specific instances. Subdivision planning requires adequate surface drainage away from buildings. The drainage plan might be designed in a manner that will cause temporary ponding by using grades which will create reduced water velocity. This can also reduce erosion.

The planner and engineer should ensure that the neighborhood does not have clays or shales underlying the surface which could affect building foundations due to excessive swell pressures. In such cases, water should not be ponded or percolated into the ground except in preselected locations. Use of subsurface drains at a shallow depth is recommended.

As an alternate to curbs and gutters, grassed depressions with a sub-surface drain can sometimes be used to limit the effects of urbanization. In the case of planned development with integral open space, such depressions could be used as the primary means for transporting runoff. Also, storage may be augmented by providing controls (weirs, checks, etc.) along such depressions. In effect, a series of linear reservoirs can be created, thereby creating storage volume. Depending upon the extent of such controls and the volume provided, storage can be obtained along with possible increased infiltration of the stored runoff.

Open Space Storage

Storage can be combined with open space and recreation areas. Open space areas can be utilized for the temporary detention of the storm runoff with a minimum effect on their primary function [6].

Recreational areas, such as soccer or football fields, generally have a substantial area of grass cover which often has a good infiltration rate. Storm runoff from such fields is generally minimal. However, in some cases where fields are underdrained, additional precautions will have to be under-taken to prevent short-circuiting and unplanned loss of storage. The mul-tiple use of such recreational fields can be made by providing for the ponding of runoff from adjacent areas.

Parks create little runoff of their own, but provide excellent storage potential. Using parks as storage areas can reduce the total urban system cost by combining capital requirements and maintenance requirements into multiple-purpose facilities.

If properly planned and constructed, utilization of parks for storage will cause little additional maintenance costs due to the storm drainage function and will often be nonconflicting with park purposes. If a permanent conser-vation pond is provided in the storage plan, recreational opportunities will be expanded. Also, a water supply could be developed for park irrigation.

Retention Reservoirs

Retention reservoirs in a catchment generally are major storage facilities. They are located in the valleys and have the ability to regulate the streamflow. Because they have permanent conservation ponds or lakes, they can be inte-grated into the system of metropolitan parks or other large open areas. Water-oriented recreational features can be incorporated into the planning of such reservoirs.

Detention Reservoirs

Detention reservoirs generally are located on streams. However, they are frequently located above the reaches where there is a continuous flow. Thus, they do not have permanent ponds and do not provide opportunities for water-oriented recreation. They can be designed as integral parts of a park and open space plan.

Gravel Pits and Quarries

Gravel pits and quarries can be designed to provide significant flood storage. Such storage is off-channel. A side-channel spillway can be used to permit the peak of the hydrograph to spill into the storage area. Outflow from such storage areas generally must be pumped.

10.3 IMPLEMENTATION

Storage planning and implementation are particularly challenging to the drainage planner and engineer. This is because storage deals with land use planning and some modification of traditional methods of site and subdivision scheme layout. Further, it is a means of storing water in a widely dispersed manner; thus, normal concepts of operating and maintenance do not apply.

Additionally, the question often arises as to whether or not storage of stormwater in some portion of a basin will actually cause an increase in the downstream flood peak as the delayed runoff coincides with a later arriving flood peak from a different basin. This argument may lead to a conclusion that runoff from the upper portion of the basin should be detained while that in the lower portion should be allowed to flow off as quickly as possible. The complexities of the total drainage system, however, would indicate that slowing of runoff from the whole basin is preferable. First, storms do not occur in a consistent pattern geographically nor temporally. Secondly, much of the total drainage dollar is spent upstream for storm drains and artifical drains in the minor tributaries. Erosion of channels in the small tributaries is a problem because generally the smaller the basin, the greater is the impact of urbanization on peak runoff magnitude. Finally, in a flood a large portion of damage occurs in the neighborhood which usually goes unreported.

Land Opportunities

Open space and recreational areas require land. To be most effective, the land should be where the people are; however, this land is also more costly. Storage areas, to be most effective, should be where the development is. It is this common factor which ties together the objectives of the urban planner, recreationalist, and drainage engineer.

Construction Opportunities

Many opportunities exist for the construction of storage facilities by simple modification of the existing sites and facilities. Frequent opportunities are presented by road embankments, which by adaption of the culvert or by the addition of an inlet structure, and the protection of the road embankment from washout, can result in a storage basin. Other opportunities include school grounds and sports fields adjacent to drainageways that are particularly suitable to storage by the addition of a berm and the construction of outlet works for draining the facility.

REFERENCES

1. *Urban Hydrology for Small Watersheds*, Technical Release No. 55, U.S. Department of Agriculture, Engineering Division, Soil Conservation Service, January 1955.

2.   *Airport Drainage*, Federal Aviation Agency, Washington, D.C., 1966.

3.   *Investigation of Porous Pavements for Urban Runoff Control*, Edmund Thelen, Wilford C. Grover, Arnold J. Hoiberg, and Thomas I. Haigh, 11034 DUY, U.S. Environmental Protection Agency, Office of Research and Monitoring, Washington, D.C., March 1972.

4.   *Lakes and Ponds*, Joachim Tourbier and Richard Westmacott, Technical Bulletin 72, Urban Land Institute, Washington, D.C., 1976.

5.   *Living with a River in Surburbia*, Bauer Engineering, Inc., a report to the Forest Preserve District of DuPage County, Ill., 1974.

6.   *Residential Storm Water Management*, Urban Land Institute, Washington, D.C., American Society of Civil Engineers, New York, National Association of Home Builders, Washington, D.C., 1975.

7.   "Storm Water Detention in Urban Areas," Eugene J. Daily, abstract of paper presented at the Eighth Annual National Highway Conference held at Atlanta, Ga., *Public Works*, Ridgewood, N.J., January 1961.

8.   *Water-Resources Engineering*, Ray K. Linsley and Joseph B. Franzini, McGraw-Hill, New York, 1964.

11
Water Quality

11.1 INTRODUCTION

Storm and snowmelt runoff from the streets, parking lots, roofs, and other types of land cover carries with it pollutants. If the runoff is routed directly to public waterways and lakes, so are the pollutants. A runoff management plan which ignores pollutant sources is incomplete, as the plan may not be amenable to incorporating such pollution control in the future.

The first pollution abatement activities were focused on the most obvious pollution source—water-borne sanitary sewage. Now good secondary treatment for all sanitary wastes is an accepted minimum standard. While considering upgrading sanitary sewage treatment above this present norm, it is rational to consider urban runoff water pollution which might now be relatively significant.

Pollutants from urban runoffs are similar to those found in ordinary municipal sewage [17]. However, quantities vary more widely with time, and tend to produce shock loadings to receiving waters. A typical urban storm runoff pollutant load from 2.5 mm (0.1 in.) of rainfall could be expected to contain more than 10,000 times as much settleable and suspended solids in an hour than would effluent from the secondary sewage plant serving the same urban area. Five-day biochemical oxygen demand (BOD) would be about 50 times, coliform bacteria nearly 900 times, and phosphorus 180 times as much as that contained in secondary sewage effluent from the same urban area during the same runoff period of an hour or two.

General Quality of Surface Runoff

Table 11.1 illustrates the estimated general quality of surface runoff. The planner of a runoff program should consider the effects and manage these constitutents as they pass through and accumulate in portions of the system.

Of particular importance are solids, especially those caused by erosion. These can have a major effect on the useful life of a man-made storage facility. These values can range from 1100 to 110,000 kg/(mi^2)(year) [10 to 1000 lb/(acre)(year)] from agricultural areas and are potentially higher from disturbed developing areas. Technical Releases Nos. 12, 32, and 51 of the Soil Conservation Service present detailed calculation methods for estimation of soil erosion [1-3].

Table 11.1 General Quality of Surface Runoff from Nonpoint Discharges

Item:	Benson [13]	McCarl [14]	Weibel et al. [15]	Weidner et al. [16]
Land use:	Farmland	Farmland	Residential	Research plots and apple orchard
BOD, mg/liter	5-30	3-15	2-84	3-8.4
BOD, lb/(acre)(year)	—	6.9	33	3.7-120
kg/(km^2)(year)	—	770	3,700	414-13,500
COD, mg/liter	50-360	70-780	20-610	40-68
COD, lb/(acre)(year)	—	246	240	27.8-1,300
kg/(km^2)(year)	—	27,600	26,900	3,100-146,000
Solids, mg/liter	90-5,000 SS	180-6,000 SS	5-1,200 SS	500-575 TS
Solids, lb/(acre)(year)	—	2,040 SS	730 SS	185-13,200 TS
kg/(km^2)(year)	—	229,000 SS	81,800 SS	20,700-1,480,000 TS
Total phosphorus, mg/liter	0.26-2.4 (Soluble P)	0.04-0.60	0-1.4 (Soluble P)	0.42-0.98
Total phosphorus, lb/(acre)(year)	—	0.07	0.8	0.36-9.0
kq/(km^2)(year)	—	7	90	40-1,000
Org. N + NH$_3$, mg/liter	1.3-20.3	2.8-17	0.1-7	—
Total nitrogen, mg/liter	—	12.9-33.2	0.2-9	4.9-9.0
Total nitrogen, lb/(acre)(year)	—	8.4	8.9	0.8-237

Key: BOD = biological oxygen demand; COD = chemical oxygen demand; SS = suspended solids; TS = total solids.
Source: Reproduced from Ref. 4.

Runoff Pollution from Urban Areas

Urban runoff pollution loads have been studied and measured in many places so that estimates based upon present state-of-the-art can be made for other communities. This extrapolation is made because certain similarities exist between cities as to patterns of auto use, street and housing development, and urban cleanliness. Presented here are typical urban characteristics with related pollution loads from runoff. Comparisons are made to sanitary sewage loads that put the runoff pollution loads into more readily understandable perspective.

The URS Research Company in *Water Pollution Aspects of Street Surface Contaminants* [5] and *Toxic Materials Analysis of Street Surface Contaminants* [6] made the following projections for a hypothetical city:

Assumed population unit	100,000 people
Total land area	57 mi^2 (14,000 acres)
Land use distribution	
Residential	75%
Commercial	5%
Industrial	20%
Total street lengths	640 curb km (400 curb mi)
Sanitary sewage flow	45 million liters/day (12 million gal/day)
Rainfall	2.5 mm (0.1 in.) having short-duration intensity of 12.7 mm/hr (0.5 in./hr)

Table 11.2 gives typical pollutant loads during a 2.5-mm (0.1-in.) rainfall event based on measured pollution data from numerous communities (San Jose, Phoenix, Milwaukee, Baltimore, Atlanta, Tulsa, Seattle) and the comparison to raw sanitary sewage.

The relative impact of surface runoff pollution during a rainfall event compared to the effluent of a secondary sewage treatment is greater.

Another report, *Impact of Various Metals on the Aquatic Environment* [7], indicates that the loading of copper, lead, and zinc in soft water can be toxic to aquatic organisms. Studies by the Colorado Department of Game, Fish, and Parks indicate that remarkably low concentrations of certain heavy metals cause significant problems with fish, the effect being related to water hardness.

The data illustrate that storm runoff from developed urban areas represents a significant source of water pollution. Table 11.2 indicates the levels of pollutant discharges relating to short-term "shock"-type loadings. Table 11.3 gives a comparison of average daily stormwater pollution load and sewage loads for Tulsa, Oklahoma. This table shows that the loads are significant even on an average basis. On a total stream pollution basis, however, the relationships are more complex. For example, runoff pollution would not be expected to occur during the defined critical periods, nor is the duration of high pollutant concentration continual during long runoff periods.

In summary, the above discussion illustrates the pollution problem associated with large urban areas. The water quality testing program in smaller areas illustrates that the problem is similar [8,9].

Runoff pollutants are provided by a myriad of urban activities such as development, traffic, air pollution, and human actions [18]. Construction activities, littering, animals, poor maintenance of drainageways, material storage, residential and commercial site deficiencies, land uses, and indiscriminate dumping practices are all contributing factors.

Table 11.2 Comparison of Pollutional Loads from Hypothetical City: Street Runoff vs. Raw Sanitary Sewage

	Contaminant loads on receiving waters street surface runoff		Raw sanitary sewage			Street/sewage ratio
	(kg/hr)	(lb/hr)	(mg/liter)	(kg/hr)	(lb/hr)	
Settleable and suspended solids[a]	254,000	560,000	300	590	1,300	430
BOD5[a]	25,400	5,600	250	500	1,100	5.1
COD[a]	5,900	13,000	270	540	1,200	11
Total coliform	40×10^{12}		250×10^6	4.6×10^{14}		0.0087
Bacteria	Organisms/hr		Organisms/(liter organisms)(hr)			
Kjeldahl nitrogen[a]	400	880	50	100	210	4.2
Phosphates	200	440	12	23	50	8.8
Zinc	120	260	0.20	0.38	0.84	310
Copper	40	80	0.04	0.08	0.17	470
Lead	100	230	0.03	0.06	0.13	1,800
Nickel	9	20	0.01	0.02	0.042	480
Mercury	13	29	0.07	0.12	0.27	110
Chromium	20	44	0.04	0.08	0.17	260

[a] Weighted average by land use, all others from numerical mean.

Key: BOD5 = biochemical oxygen demand, 5 days; COD = chemical oxygen demand.

Source: Reproduced from Ref. 5.

Table 11.3 Estimated Daily Load of Pollutants Entering the Area Receiving Streams for Tulsa, Oklahoma

Parameter	Average daily storm water pollution load, lb	1968 Average daily load from sewage treatment facilities, lb	Total load	Percentage contribution of storm water to total load
BOD	4,455	19,370	23,825	20
COD	30,803	67,180	97,983	31
Suspended solids	107,200	18,400	125,600	85
Organic Kjeldahl nitrogen	355	760	1,115	31
Soluble orthophosphate	469	11,020	11,489	4

Key: BOD = biological oxygen demand; COD = chemical oxygen demand.
Source: Reproduced from Ref 10.

The following list gives a detailed breakdown of pollutants found on typical streets in various degrees [6]:

A. Large-sized/biologically insignificant
    1. Bulk cellulosic matter
        a. Tree limbs, twigs, leaves, shrubs
        b. Lumber
        c. Paper
        d. Cotton materials
        e. Rayon
        f. Cellophane
    2. Bulk metals and alloys of construction and containerization
        a. Steel
        b. Iron
        c. Aluminum
        d. Magnesium
        e. Copper and bronze
        f. Zinc
        g. Tin
    3. Fabric, packaging, and construction plastics
    4. Natural processed animal fibers

B. Variable-sized/biologically insignificant
    1. Soil conditioners
    2. Basic soil constituents
    3. Inorganic dustfalls from air pollutants

C. Variable-sized/biologically nutritive/water-soluble
    1. Natural and compounded fertilizers
        a. Nitrogen compounds (ammonium, nitrate, urea, cyanates, etc.)
        b. Phosphates
        c. Potassium compounds
        d. Secondary growth elements (Ca, Mg, Fe, Cu, Zn, Mn, B, Mo, S)
    2. Deicing compounds
        a. Sodium hexametaphosphate
        b. Urea
        c. Ammonium nitrate
        d. Potassium pyrophosphate
    3. Soluble air pollutants
        a. Sulfur oxides (as SO_4)
        b. Nitrogen oxides (as NO_3)
        c. Ash
    4. Phosphate-based detergents
    5. Lawn and garden ash

D. Variable-sized solids or solutions/biologically inhibiting/water-soluble
    1. Deicing compounds
        a. Sodium chloride
        b. Calcium chloride
        c. Ferric ferrocyanide
        d. Sodium ferrocyanide
        e. Sodium chromate
    2. Air pollutants
        a. Carbon monoxide
        b. Sulfides, sulfites

        c. Nitrites
        d. Ozone
    3. Antifreeze compounds
        a. Diacetone alcohol
        b. Methanol
        c. Ethylene glycol
    4. Roadway hydrocarbons—some highly oxygenated bitumens
    5. Water-base paint solutions

E. Variable-sized/immiscible or suspendable/biologically inhibiting/
    water-insoluble
    1. Vehicular and roadway hydrocarbons
        a. Oils
        b. Greases
        c. Tetraethyl lead and decomposition products
        d. Bitumens
    2. Hydraulic fluids
        a. Propylene glycol diricinoleate
        b. Trinitrobutylamine
    3. Water-insoluble air pollutants—hydrocarbons
    4. Pesticide/herbicide carriers

F. Variable-sized solids or solutions/biologically toxic/water-soluble
    1. Common pesticides, herbicides, etc.
        a. Arsenic (acetoarsenites, arsenites, arsenates)
        b. Copper (arsenites etc.)
        c. Lead (arsenites etc.)
        d. Thallium compounds
        e. Chloropicrin
        f. Dinitro-o-cresol
        g. Furfural
        h. Malathion
        i. Nicotine
        j. Phenol

G. Variable-sized solids, liquids, or suspensions/biologically
    toxic/water-insoluble
    1. Common pesticides, herbicides, etc.
        a. Benzene hexachloride
        b. Chlordane
        c. Dichlorodiphenyltrichloethane (DDT)
        d. Dichloroethylene
        e. Dichloroethyl ether
        f. 2-4-Dichlorophenoxycetic acid (2,4-D)
        g. Dinitro-o-cresol
        h. Methoxychlor
        i. Parathion
        j. Tetramethylthiuram disulfide
        k. Toxaphene
        l. Trichloroethylene
        m. Dichlorobenzenes (ortho and para)
        n. Pyrethrins
        o. Aldrin
        p. Dieldrin
        q. Organo-mercury compounds

   H. Variable-sized culture media/biologically active/ water-suspendable
       life forms
      1. Animal excretions
         a. Fecal coliforms
         b. Fecal streptococci
         c. Biological nutrient source
      2. Human excretions
         a. Fecal coliforms
         b. Fecal streptococci
         c. Biological nutrient source
      3. Dead animals
         a. Fecal coliforms
         b. Nonfecal coliforms
         c. Fecal streptococci
         d. Biological nutrient source
      4. Vegetation–biological nutrient source
      5. Food wastes–biological nutrient source
      6. Soil–biological nutrient source

11.2 RUNOFF WATER QUALITY SIMULATION

As discussed in Chap. 5 on runoff hydrology, computer models exist which
can be used to estimate water quality characteristics of runoff.

These quality models are usually based upon a washoff algorithm and/or
the universal soil loss equation for depicting erosion. The reader should keep
in mind that this science is in a relatively early stage of development; therefore, predictive results should be tempered with good judgment.

Initial Theory of Washoff of Pollutants

As the start of a rainfall event, the amount of a particular pollutant on surfaces which produce runoff (both impervious and pervious) will be P_0.
Assuming that the weight of pollutant washed off in any time interval dt is
proportional to the weight remaining on the ground P, the first-order differential equation is:

$$\frac{-dP}{dt} = kP \tag{11.1}$$

which integrates to:

$$P_0 - P = P_0(1 - e^{-kt}) \tag{11.2}$$

in which $P_0 - P$ equals the weight washed away in the time t.

In order to determine k, it was assumed that k would vary in direct proportion to the rate of runoff r, or $K = br$. To determine b it was assumed
that a uniform runoff of 12.7 mm/hr (1/2 in./hr) would wash away 90 percent
of the pollutant in 1 hr. This leads to the equation:

$$P_0 - P = P_0(1 - e^{-0.18rt}) \qquad \left[P_0 - P = P_0 (1 - e^{-kt}) \right] \tag{11.3}$$

              (metric)   (British)

where r = runoff rate, mm/hr (in./hr)
    t = time interval, hr

The assumption of 12.7 mm/hour for b has proven satisfactory in all test
applications to date for urban areas.

In using this equation, a uniform time step t is selected, values of r are
determined from the inlet hydrograph, and the equation is applied successive-
ly, the value of P determined at the end of the nth interval becoming the
value of P_0 at the beginning of the (n + 1) th interval. The value of r equals:

$$r = 0.5 \, (r_n + r_{n+1}) \tag{11.4}$$

Various corrective factors are used in the different methods to derive
final results. These factors consider such parameters as the actual availability of pollutants, streetsweeping effectiveness, and characteristics of the
drainage system of interest [10].

Estimation of Erosion

A commonly used method for the calculation of soil loss is the *universal soil-
loss equation*. The actual amount of this eroded soil which travels in the
waterways is only a portion of this. The ratio of how much of the eroded soil
enters the waterway is known as the sediment delivery ratio [1-3].

The basic soil loss equation is:

A= RKLSCP

where A = soil loss per unit area
       R = rainfall factor
       K = soil erodibility factor
       L = slope length factor
       S = slope gradient factor
       C = cropping management factor
       P = erosion control practice factor

Reference to Models

The various models mentioned in Chap. 3 give further details. Note that
these models can be used to assess the effects of various water quality management techniques to be discussed in the following section.

11.3 POSSIBLE COMPONENTS FOR RUNOFF POLLUTION
      TREATMENT

The flood control measures and water quality control measures of drainage
system design are closely interrelated. The following paragraphs will present
various alternatives to the water quality problem from which a program can be
formulated.

The controls and treatment systems for urban surface runoff management
in any location could be the combination of several approaches that are shown
in Fig. 11.1. These are described below.

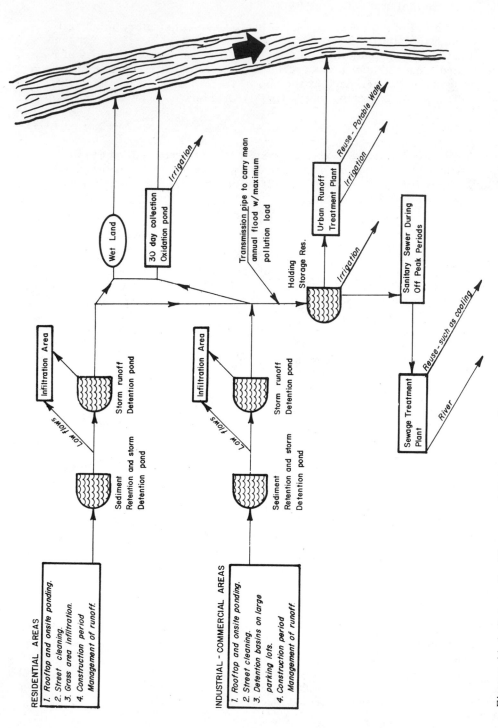

Figure 11.1 Schematic Diagram of Urban Runoff Management Opportunities.

Rooftop and On-Site Ponding

The concept of rooftop and on-site ponding is to regulate quantities (up to 100-year frequency flood flows) to historical conditions. These facilities would also help to limit the pollution washoff rate.

Street Cleaning

Street cleaning, where surface pollutants are removed from the street and sent to a solid waste disposal site on a *frequent basis* (daily or semidaily), helps curtail total and shock loads that are received by the stream or river. Street washing without such controlled disposal does no more than wash pollutants to the stream, providing little benefit. Vacuum sweeping has been noted to be more effective in comparison to mechanical sweeping.

On-Site Grass Area Infiltration

On-site grass area infiltration provides treatment by allowing the pollutants and water to be absorbed, infiltrated, and not discharged directly to waterways.

Detention Basins on Large Parking Lots

Detention basins on large parking lots are beneficial from both flow and pollution standpoints because the sudden flushing effect is prevented and much of the pollution will not be carried off. In fact, much of the pollution will stay in the same vicinity and the latter can be picked up by cleaning measures. The design can provide for a normally dry surface like a typical lot, filling only in severe rainstorm events.

Construction Period Management of Runoff

There are a multitude of building code items and site development regulations that can be initiated to require measures before groundbreaking for the site, such as:

1. Diversion terraces for upstream flows
2. Velocity controls for existing or proposed swales
3. Soil stabilization
4. Collection of runoff with discharge to onsite sedimentation/treatment ponds or controlled flow spreading in absorptive vegetated areas
5. Sedimentation basins that have volume/surface area constraints for areas of development
6. Specialized controls and filtering devices in large earthmoving and construction projects

The reader is referred to EPA study R2-72-015, *Guidelines for Erosion and Sediment Control Planning and Implementation* [11]. The necessary measures should be anticipated and implemented before construction of any size development or improvements.

Sediment Retention and Storm Detention Ponds

Sediment retention and storm detention ponds can be used to regulate flows out of an entire subbasin, providing many benefits in terms of smaller storm sewer, channel, and/or swale improvements. Much of the suspended solids loading can be removed so that the receiving streams feel less impact. These ponds can also be used for significant storage of storm runoff to provide highly controlled flow to treatment/control facilities.

Infiltration Areas

Infiltration areas can be used to treat low and regulated flows from the retention and detention ponds. These areas can be open space floodplains and recreation areas.

Wetland

Wetland such as marshes, floodplain overbank areas, and alluvial terraces provide nature's way of filtering and treating flow before discharge to main water bodies. A controlled, shallow-flow-type design criterion will benefit water quality, especially in conjunction with many of the measures above.

Oxidation Ponds

Oxidation ponds usually will provide an adequate system for treatment and control. The concept would be to construct a pond large enough to hold the heaviest average monthly rainfall. Natural oxidation and sedimentation processes, similar to a series of natural beaver ponds, would occur. The pond would discharge to the stream via a filter berm or layer such that high-quality water would result. The pond would be periodically drained and its bottom cleaned before any toxic level of sludges and sediments could build up. Mechanical aeration may be necessary.

Interception Systems

Interception systems are used to divert nonpolluted tributary flows from areas of contamination, thus minimizing the quantities of polluted runoff flows.

Transmission Systems

Transmission systems could divert polluted runoff flows to more sophisticated treatment systems.

Holding Storage Reservoir

A holding storage reservoir would provide the volume capacity and flow regulation necessary to provide economical treatment of the highly varying flow-rates of surface runoff.

Sanitary Sewage Treatment Plants

Sewage treatment plants, in some cases, can be used to treat regulated flows from the holding storage reservoirs during off-peak hours. This would better balance flows through the plant, make efficient use of existing facilities, and essentially prevent discharge of urban runoff pollutants.

Irrigation

Irrigation requirements could be met by using waters held in the holding storage reservoir in controlled land application similar to sanitary sewage treatment methods.

Urban Runoff Treatment Plants

Urban runoff treatment plants could be built specifically to treat the polluted water held in the holding storage reservoir. The effluent could be used for irrigation, sent directly to the river, or used economically for a potable water supply along with other uses, as has been shown in one study [12].

However, the runoff management scheme must be flexible and be coordinated with the sewage facilities. The runoff management plan can designate basic ponds, swales, channels and piping systems, possible treatment schemes, and a proposed final treatment scheme with the final choice of the domestic sewage treatment facilities for the total wastewater picture.

REFERENCES

1. *Procedure—Sediment Storage Requirements for Reservoirs*, Technical Release No. 12, U.S. Department of Agriculture, Engineering Division, Soil Conservation Service, March 1959; revised January 1968.
2. *Procedure for Determining Rates of Land Damage, Land Depreciation, and Volume of Sediment Produced by Gully Erosion*, Technical Release No. 32, U.S. Department of Agriculture, Engineering Division, Soil Conservation Service, July 1966.
3. *Procedure for Computing Sheet and Rill Erosion on Project Areas*, Technical Release No. 51, U.S. Department of Agriculture, Engineering Division, Soil Conservation Service, 1972; revised September 1977.
4. *Quantification of Pollutants in Agriculture Runoff*, Environmental Protection Technology Series EPA-660/2-74-005, U.S. Environmental Protection Agency, Office of Research and Development, Washington, D.C., February 1974.
5. *Water Pollution Aspects of Street Surface Contaminants*, Environmental Protection Technology Series EPA-R2-72-081, U.S. Environmental Protection Agency, Office of Research and Development, Washington, D.C., November 1972.
6. *Toxic Materials Analysis of Street Surface Contaminants*, Environmental Protection Technology Series EPA-R2-73-283, U.S. Environmental Protection Agency, Office of Research and Development, Washington, D.C., August 1973.
7. *Impact of Various Metals on the Aquatic Environment*, Technical Report No. 2, U.S. Environmental Protection Agency, Office of Water Quality, Washington, D.C., 1971.
8. *Urban Runoff Management Plans, City of Aspen, Colorado*, Wright-McLaughlin Engineers, Denver, 1973.
9. *Runoff Management Plan, Snowmass at Aspen, Pitkin County, Colorado*, Wright-McLaughlin Engineers, Denver, 1975.
10. *Storm Water Pollution from Urban Land Activity*, Water Pollution Control Research Series 11034 FKL 07/70, U.S. Department of the Interior, Federal Water Quality Administration, Washington, D.C., 1970.

11. *Guidelines for Erosion and Sediment Control Planning and Implementation*, Environmental Protection Technology Series EPA-R2-72-015, U.S. Environmental Protection Agency, Office of Research and Development, Washington, D.C., 1972.

12. *The Beneficial Use of Storm Water*, Environmental Protection Technology Series EPA-R2-73-139, U.S. Environmental Protection Agency, Office of Research and Development, Washington, D.C., 1973.

13. "The Quality of Surface Runoff from a Farmland Area in South Dakota During 1969," R. D. Benson, M.S. Thesis, S.D. State Univ., Brookings, S.D., 1970.

14. "Quality and Quantity of Surface Runoff from a Cropland Area in South Dakota During 1970," T. A. McCarl, M.S. Thesis, S. D. State Univ., Brookings, S.D., 1971.

15. "Rural Runoff as a Factor in Stream Pollution," Journal Water Pollution Control Federation, 41, 377, 1969.

16. "Urban Land Runoff as a Factor in Stream Pollution," Journal Water Pollution Control Federation, 36, 914, 1964.

17. *Urban Runoff Characteristics*, 11024 DQU, Environmental Protection Agency, Water Quality Office, Washington, D.C., October 1970.

18. *Water Pollution Aspects of Urban Runoff*, Federal Water Pollution Control Administration, Washington, D.C., March 1969.

Index

About the Authors

JOHN R. SHEAFFER is President of Sheaffer & Roland, Inc. Dr. Sheaffer is a noted authority on flood proofing and wastewater recycling, and has published over 35 papers in these and related fields. He received the M.S. and Ph.D. degrees in geography from the University of Chicago. Dr. Sheaffer is a member of the American Clean Water Association, American Association for the Advancement of Science, Science Advisory Board of the International Joint Commission, and Urban Planning Development Council for the American Society of Civil Engineers.

KENNETH R. WRIGHT is the President of Wright Water Engineers and a Managing Partner of Wright-McLaughlin Engineers. His special interests include surface and groundwater supplies, water allocation, and flood management. He is a member of the Colorado State Board of Registration for Professional Engineers and Professional Land Surveyors, as well as American Society of Civil Engineers, National Society of Professional Engineers, International Commission on Irrigation and Drainage, and American Water Works Association.

WILLIAM C. TAGGART is a partner in the firm of Wright McLaughlin Engineers. His research interests include optimization of runoff storage systems, sediment transport, runoff computer modeling, and water system intakes. Mr. Taggart is a member of the American Society of Civil Engineers, National Society of Professional Engineers, and Colorado Society of Professional Engineers.

RUTH M. WRIGHT is an attorney, specializing in drainage and flood control law, and a State Representative in the Colorado State Legislature. She has authored model floodplain regulations, reports on the storm water laws of several states, and papers on nonstructural flood control. Mrs. Wright is a member of the American and Colorado Bar Associations, and the U.S. Committee on Irrigation, Drainage, and Flood Control.